RobotBASIC Robots for Beginners

Building Inexpensive **RROS**-Based Robots
(**RROS**: **R**obotBASIC **R**obot **O**perating **S**ystem)

John Blankenship

Copyright © 2017 by
John Blankenship

All rights reserved. No part of this document may be reproduced or transmitted in any form or by any means, electronic or mechanical, including photocopying, recording, or by any information storage or retrieval system without the prior written permission of the copyright owner.

Trademarked names may appear in this book. Rather than use a trademark symbol with every occurrence of a trademarked name, we use the names only in an editorial fashion and to the benefit of the trademark owner, with no intention of infringement of the trademark.

Images of proprietary devices and sensors are reproduced with the permission of the manufacturing companies.

The information and projects in this book, including the RROS chip itself are provided for hobby and educational purposes without warranty. Although care has been taken in the preparation of this book and all associated software, neither the authors or publishers shall have any liability to any person or entity with respect to any special, incidental or consequential loss of profit, downtime, goodwill, or damage to or replacement of equipment or property, or damage to persons, health or life, or any costs of recovering, reprogramming, or reproducing any data, caused or alleged to be caused directly or indirectly by the information contained in this book or its associated web site.

Table of Contents

Chapter	Title	Page
1	Why a RROS Chip	1
2	Building the Chassis	11
3	Motor Control Electronics	15
4	Controlling the Robot's Movements	21
5	Adding a Ranging Sensor	27
6	Avoiding Objects While Roaming	33
7	Adding a Turret	45
8	Programming the Robot to Follow a Wall	49
9	Adding a Line Sensor	61
10	Avoiding Objects on a Line	71
11	What's Next?	75
Appendix A	Breadboarding	81

Chapter 1
Why a RROS Chip

Not long ago, it was very difficult to build a hobby robot capable of interesting behaviors because you had to design and build nearly everything yourself. Today, robotics can be a fantastic hobby for nearly anyone because technology has advanced to the point that most of the complicated things you need can easily be purchased for reasonable prices. Let's look at some simple examples to help clarify this point.

Humans and Robots Need Sensors
Humans have many senses (vision and touch, for example) to help us interact with our environment. Robots also need sensors so they can avoid objects, navigate a house, accomplish goals, etc. A common sensor used on hobby robots is the Ping ultrasonic sensor (see Figure 1.1) available from Parallax (www.Parallax.com).   This sensor allows your robot to detect objects in its path and measure the distance to them.

Figure 1.1: A Ping Ultrasonic Sensor

The Ping module has two round ultrasonic transducers that act like a speaker and a microphone. The speaker is used to produce a short burst of sound that is too high for humans to hear. After the sound is generated, your robot must wait for the sound waves to travel to some nearby object and bounce back to be detected by the microphone. Sound travels about one foot every millisecond (1/1000 of a second) so we can calculate the distance to the object by measuring how long it takes for the sound to return to the microphone.

In the past, a hobbyist would have to build an oscillator to generate the correct frequency for the sound wave as well as a special filter that could separate the desired sound wave from other noise in the room. As you might expect, building such circuits required a reasonable knowledge of electronics as well as skills such as soldering and using instruments such as a meter and oscilloscope.

Today, all the necessary electronics is integrated into the Ping module. Using ports from the robot's computer, we can tell the module to produce the burst of sound, and it can tell us if and when it detects the sound wave returning from objects within the robot's environment. Having such electronics already built and ready to use is fantastic – but, we still need to measure the time the sound takes to complete its journey. Such programming generally requires a language such as C or assembly language because their execution speed makes it possible to time events down to milliseconds, or even microseconds (millions of a second). If you are like many novice hobbyists, you might not even know what languages like C and assembly are. You almost certainly aren't prepared to write a program to measure the time the sound travels and translate that time into inches. Don't worry, there is an easy way to overcome these obstacles.

Controlling Motors
Let's look at one more example though, before a solution is revealed. Small mobile robots often have wheels driven by DC motors. Your robot's computer will need special, high-current circuitry that allows low-current signals from a program to turn the motors ON and OFF. But, just ON and OFF is not enough because that would be like driving your car by always pressing the gas and brake pedals to the floor. Just as with a car, your robot should start moving slowly at first, and build its speed over time to prevent jerky starts and stops. The question becomes, how can we control the motor's speed if all we can do is turn it ON and OFF. The basic answer is to only turn the motor ON for a given percentage of the time. The principle is simple. If we only turn the motor ON 50% of the time it should run at 50% of its full speed. Turn it ON 90% of the time and it should run at 90% of its full speed. But this will not work if the motor is turned ON and OFF at a slow rate.

Imagine repeatedly turning the motor ON for one second, then OFF for one second. The motor would run at full speed for one second then stop for one second, over and over. Obviously, slowly turning the motor ON and OFF does not produce the solution we want. But, if the motor is only turned ON for a millisecond and then OFF for a millisecond (over and over), the motor will indeed run at about half its full speed. If we continually turn the motor ON for two milliseconds and OFF for one, then the motor will run faster – at approximately 2/3 of its full speed. Controlling the motor's speed quickly like this also requires a programming language like C or assembly that can pulse the motors (ON and OFF) very quickly to control their speed.

Furthermore, the percentage of ON-time to OFF-time should be easily controllable so your programs can vary the robot's speed. And, the speed of each wheel (on a two wheeled robot) should be handled independently to ensure the robot can move in a straight line and make slow or fast turns when needed. Finally, as mentioned earlier, the percentage of ON-time needs to start very low and increase slowly until the motor achieves the desired speed to prevent jerky movements.

It is complexities such as these that make it difficult for many newcomers to robotics. It can often take weeks or even months to learn just the basic principles for controlling motors and reading sensors. During that time it is easy to become bored with robotics because in spite of all your hard work the robot you are building really can't do anything exciting.

Making it Easy

So, what is the solution? I helped develop the programming language RobotBASIC with Samuel Mishal (easily the greatest programmer I have ever known). We made it extremely easy to use even though it is a powerful tool for controlling robots, writing video games, and learning about programming in general. RobotBASIC runs on a standard PC, not on Apple products or Android tablets and phones.

I also personally developed the RROS system which is an inexpensive computer chip that can handle the hardware interface as well as all the complex, low-level, programming tasks needed for controlling a robot's motors and reading its sensors.

Using RobotBASIC to control a RROS-based robot is one of the easiest way to build a *real* robot (one capable of interacting with its environment). Using this system, those new to robotics can often have a robot built and doing something interesting in a few hours. Even many advanced hobbyists love the system because it allows them to focus all their attention on sophisticated subjects such as in-home navigation, vision, artificial intelligence, etc. because they don't have to worry about the hardware and software needed for low-level motor control and sensor interrogation. Information on purchasing the RROS chip can be found at www.RobotBASIC.org.

A System You Can Grow With

It is important to realize that even though this book shows you how to quickly build inexpensive, easy-to-control, robots, that you can also use this system to control sophisticated robots like Arlo, shown in Figure 1.2. Arlo uses the RROS system for his mobility and main sensors. Everything, including his many additional capabilities, are explained in the book **Arlo: The Robot You've Always Wanted**, available on www.Amazon.com. The important point is that everything you learn about beginner-bots in this book can be applied to advanced robots like Arlo as your knowledge and skills increase.

RobotBASIC

Let's look at a simple example to see how easy it can be to control a robot with RobotBASIC which will be used throughout this book. RobotBASIC is a totally FREE language available from www.RobotBASIC.org. It is free for hobbyists, schools, teachers, students… everyone! And it really is free. Unlike many "free" things, the version you download is the *full* language and it does *not* time out after 30 days or force you to upgrade to get something you can really use.

After downloading the zip file, just unzip it and you are ready. No installation is required. Just open the RobotBASIC folder and run the RobotBASIC.exe application. Note: The download includes many demonstration programs and the 300 page help file that you can explore in detail on your own. When first run, RobotBASIC will ask you to accept the user license. Read the license of course, but basically it tells you that RobotBASIC is free and that you agree not to sell it to others (although you can certainly sell programs you write with RobotBASIC).

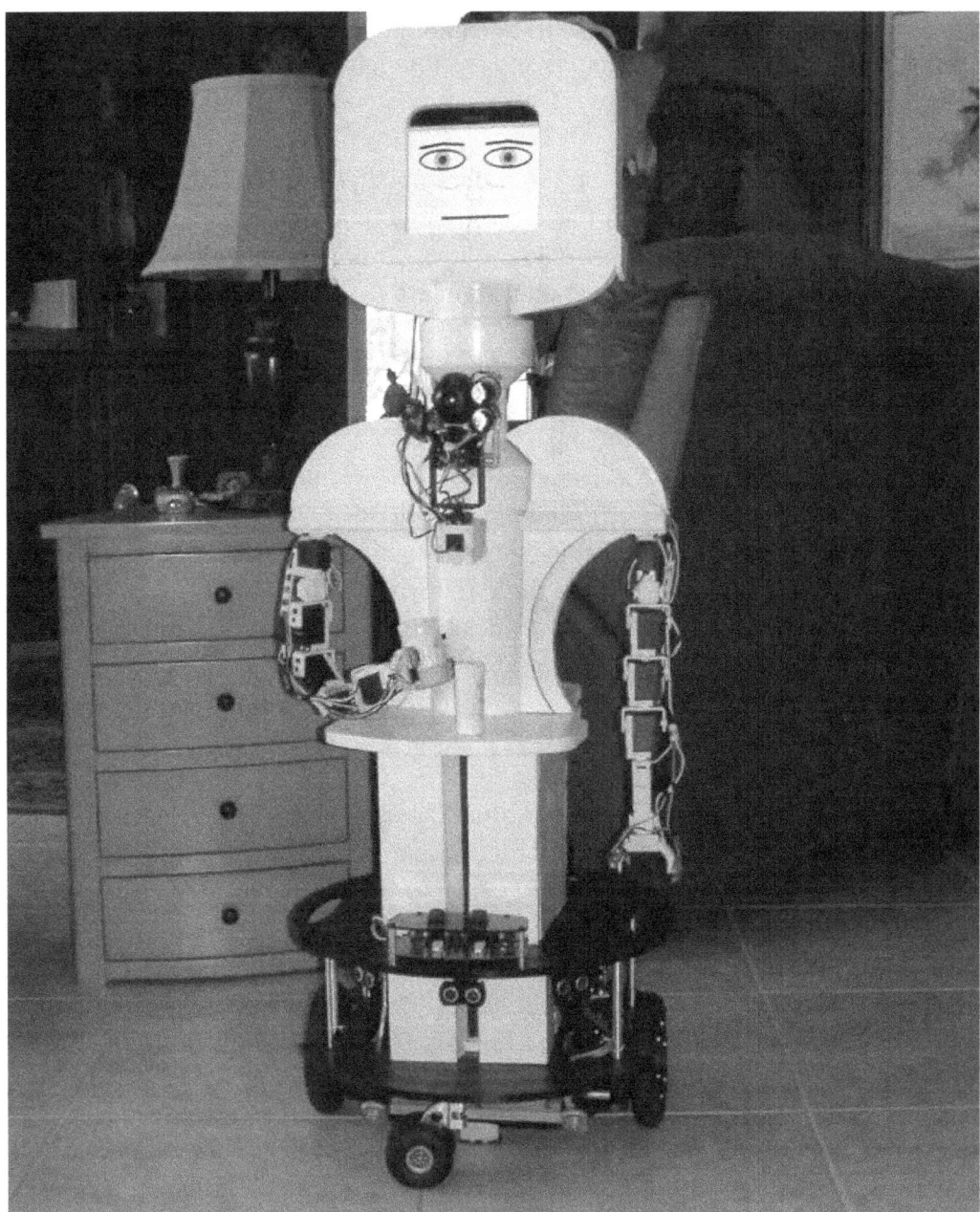

Figure 1.2: Arlo: The Robot You've Always Wanted

The RobotBASIC IDE
Figure 1.3 shows the Integrated Development Environment (IDE) for RobotBASIC. There are drop-down menus across the top with a variety of short-cut buttons below. The green triangle

button is the short-cut to run programs. Various actions will be introduced as needed throughout this book but you can always get detailed information on everything in RobotBASIC by clicking the **?** button to read the help file.

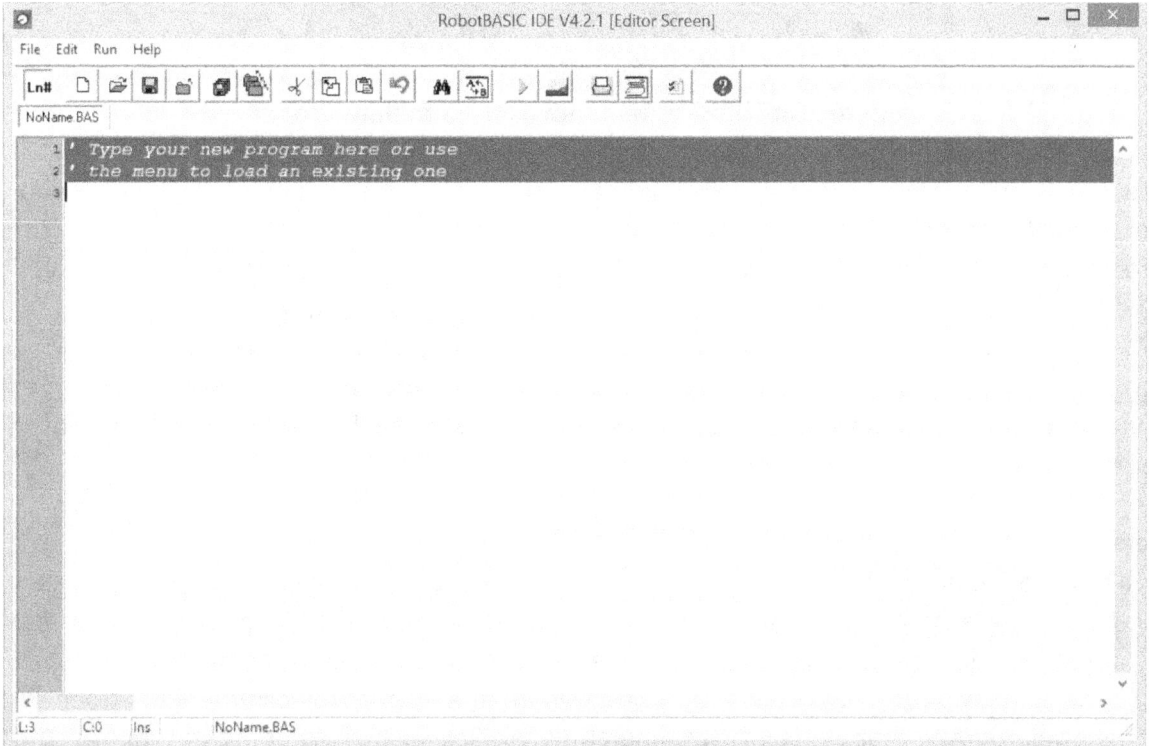

Figure 1.3: RobotBASIC's IDE

RobotBASIC's Simulated Robot

One of the great things about RobotBASIC is its integrated robot simulator. The short program in Figure 1.4 shows how easily you can control the simulation. The first line in the program creates the simulated robot by locating it at position 400,300. Since the output screen is 800 by 600, the robot is created in the center of the screen. All robot related commands in RobotBASIC begin with the letter **r**. The second line slows the robot down so it is easier to watch. The parameter is a delay time, so zero is the fastest with larger numbers producing slower movements. The third line drops a pen so that the robot leaves a trail as it moves. Leaving a trail will be used often in this book to make it easier for you to follow what the robot is doing.

The next 3 lines move the robot forward 40 units, turns it right 90°, and moves it forward again 80 units. For the simulator, the units are screen pixels. Since the robot is 40 pixels in diameter, the first move is a distance equal to the robot's diameter. The second is twice the robots diameter.

```
rLocate 400,300
rSpeed 10
rPen DOWN
rForward 40
rTurn 90
rForward 80
end
```

Figure 1.4: This program moves the simulated robot.

Once you have the programmed typed in, you can run it by clicking the green triangle. If you have typed the program in correctly, you will see the robot move as shown in Figure 1.5. Notice the output window pops up in front of the IDE. As expected, the robot is initialized in the center of the screen and faces upward (considered north to the simulator) by default.

The trail left by the robot as it moves shows that the robot initially moved a distance approximately equal to its diameter before turning right and moving twice its diameter. The simulator may look unsophisticated, but it has many exciting capabilities that will be explored in later chapters.

Errors

If you made a typing error when you entered the program, the program will not respond as expected. If, for example, you typed **rForward 4** instead of **rForward 40**, the robot would travel only a small amount on its first move. Errors like this require you to watch the robot's behavior to see if it actually does what you want it to do. When it does not, you can mentally trace through the program after watching the simulation to discover the offending statement. The important point here is that the program will do what *you tell it to do*, not necessarily what you *want it to do*.

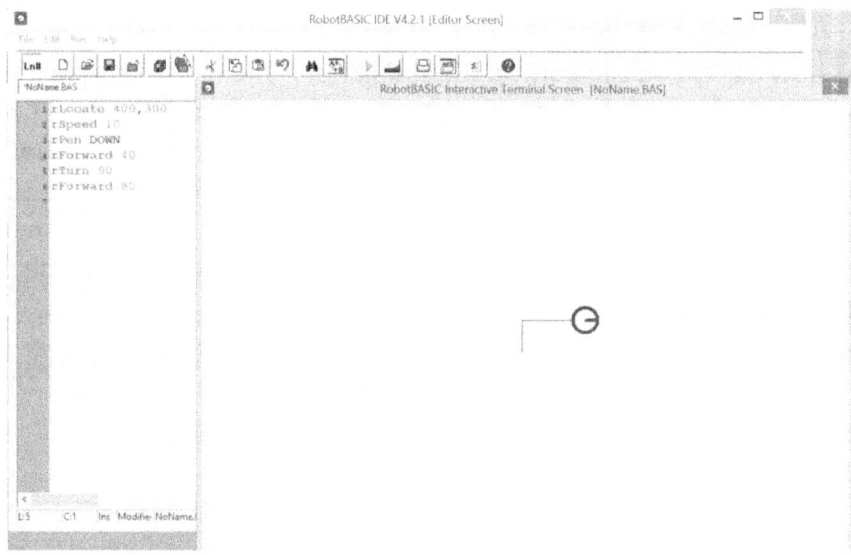

Figure 1.5: When you run the program, the robot responds as shown.

Sometimes a typo can create a situation that RobotBASIC cannot perform at all. Suppose, for example that you entered **Turn 90** instead of **rTurn 90**. RobotBASIC does not have an instruction called **Turn**, so it stops trying to run the program when it encounter that word and provides an error message (see Figure 1.6) indicating *what it thinks the problem might be*. Notice that all the statements encountered before the error were executed correctly.

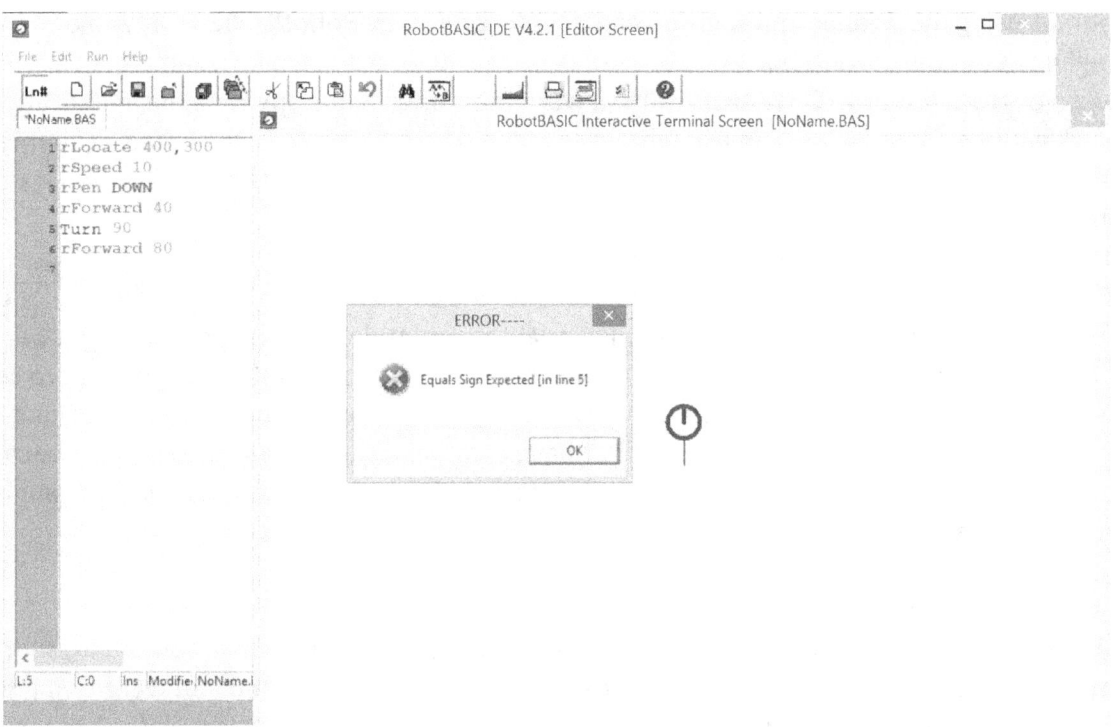

Figure 1.6: Some errors are bad enough to confuse the system.

Error Messages
The error message in Figure 1.6 is not really correct because it did not say "You misspelled **rTurn**". That would have been nice, but as you experiment with RobotBASIC (or any other programming language) you will see that *it does not know what you are trying to do*, so it just makes a *reasonable guess* at what *might* be wrong.

In this case, RobotBASIC thinks you might have been assigning a value to a variable named **Turn**. To make this clearer, let's explore variables for a moment before continuing. Assume we have a program like the one in Figure 1.7. It assigns a value of 22 to the variable **X** and a value of 3.5 to the variable **Y**. The third line in the program adds **X** and **Y** together and stores the result in the variable **dog**. The final line will print 25.5 on the output screen. Enter this program and run it to verify it works.

```
X = 22
Y = 3.5
dog = X+Y
Print dog
```

Figure 1.7

Case Sensitivity

Sometimes using the wrong capitalization can be interpreted by RobotBASIC as a typo or misspelling. Commands in RobotBASIC are *not* case sensitive. You can, for example, use **rForward** or **rforward** or **RFORWARD** – all will work the same. Since any type of capitalization will work with commands, you can use whatever makes the program easier to read. You might like the choices used throughout this book, but you can use your own preferences when it comes to commands.

Variables though, are different. As you can see from Figure 1.7, variables can be a single letter like **X** or **Y** or even **A** or **M**. You can also use words or any combination of letters like **dog** or **Cat** or even **abc**. Usually a programmer will name variables something that helps you remember what the variable is being used for. For example, instead of **dog**, it might have made more sense in this program to use a name like **sum** or **Answer**. You can use any name you want, as long as it is not a command. For example, you could not have a variable called **print** or **rForward** because these are commands in RobotBASIC. And don't forget, variables are case sensitive, so **Y** and **y** are *different* variables as are **dog** and **Dog**.

Back to Error Messages

Look back at Figure 1.6. In this case, RobotBASIC did not recognize **Turn** as a command, so it assumed it was a variable. Once it made that assumption, it then assumed you were trying to assign it a value of 90, which is why the error message indicated an equal sign was expected (RobotBASIC thought you were trying to write **Turn = 90**). It was wrong of course, but it was making a reasonable guess at what might be a way to fix the error.

Notice also in Figure 1.6 that it tells you the error is in line 5 and if you look at the line numbers to the left of the program you will see the error is indeed in line 5. If you close the error box by clicking the **OK** button, the error box and the output window will both disappear and line 5 in the editor will be highlighted. Effectively, RobotBASIC is taking you directly to the offending line so that you can correct it.

Of course, as we have seen, you cannot assume that RobotBASIC is always right when it tells you what it thinks the error is. It is up to you to look at the output from the program and the offending line and decide on your own what the error really is. Sometimes the error might not even be in the line that RobotBASIC thinks caused the error. As we proceed through this book, you will learn more about errors and how to find and correct them.

We have explored the RobotBASIC simulator because the simulated robot is very easy to control (as you have just seen). The great thing is that you can use the simulator to learn about programming a robot even before you build one. And… this is very important… once we do build a real robot using the RROS system mentioned earlier, you will be able to control the real robot using most of the commands used to control the simulator. As we proceed though this book, you will discover that the simulator is a powerful tool that helps you develop programs for controlling real robot.

Now that you know a little about RobotBASIC and the RROS system, let's move on to Chapter 2 and start building a robot.

Chapter 2
Building the Chassis

This chapter moves us toward our goal by beginning the physical construction of a small robot. For large or complex robots, purchasing a chassis, complete with motors, is often an excellent option. If this is your first robot, though, building your own can be fun and it can greatly lower your costs. Eventually most people will want something sturdier and more powerful, but it makes sense for most beginners to minimize costs until they get enough experience to know what they might want to buy. For that reason let's build the cheapest robot chassis ever.

The size and shape of your chassis will be determined by your motors. Figure 2.1 shows two great options for a first robot. The motors on the right are DC motors and wheels readily available on ebay for only a few dollars (search ebay for **robot motor wheel**). The motors on the left are *continuous rotation servomotors* from www.Parallax.com and many other sources. The DC motors are cheaper and work very well. Generally, the hardware and software needed to control DC and servomotors is totally different, but RobotBASIC's RROS chip can control either type using the same RobotBASIC commands. This means you can build your robot using either of these motor types and it can be used for all the projects in this book.

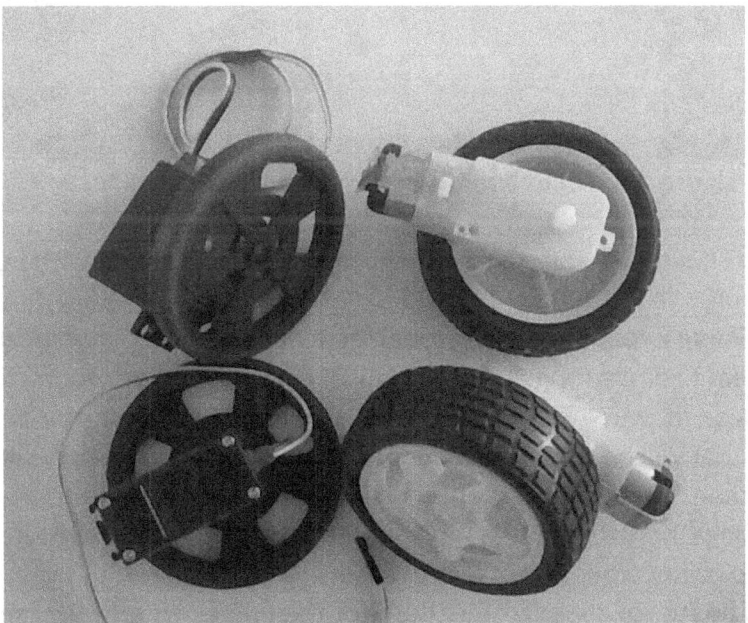

Figure 2.1: Small servo and DC motors are both excellent choices for a first robot

Once you have chosen your motors, work can begin on the chassis. If you have the proper tools and skills, you could use plastic, wood, or metal, but foam board (available at most craft stores) can be easily cut with an X-ACTO knife or razor blade and it can create an excellent base for small robots. Figure 2.2 shows how easily bases can be cut for both the DC and servomotors shown in Figure 2.3. A diameter of 6.5 to 7 inches should accommodate most battery and electronics options needed for the projects in this book. Obviously, the chassis should be large enough to meet your needs, but it should be kept as small as possible for many reasons (weight, chassis strength, maneuverability, etc.). You might want to read forward a few chapters to see what's to come before making a final decision on the size of your chassis.

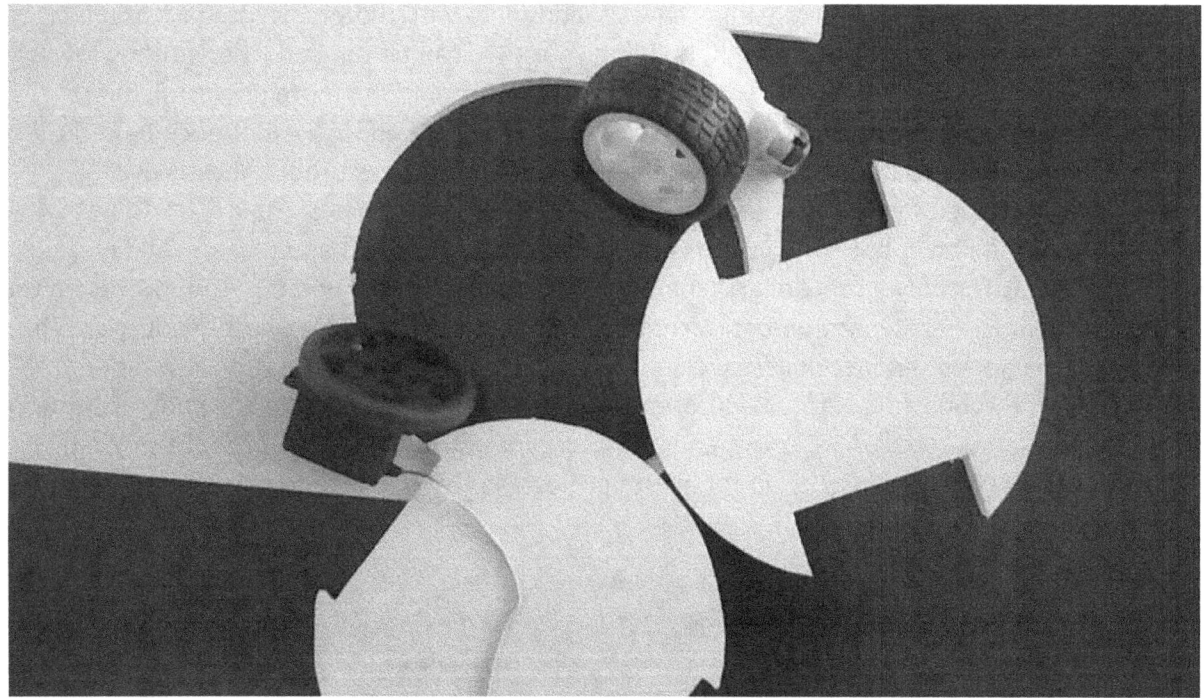

Figure 2.2: Foam board is easily cut to your specifications.

Once the base plates have been cut, the motors can easily be attached with hot glue. Such mounting is far stronger than you might imagine, yet the motors can be removed to be used in future projects. Building a robot like this is ideal for schools that might want students to build new robots every year. For a stronger motor mount, glue one or two wood strips around the edges of the motors as shown in Figure 2.3. These strips are NOT really needed, but they do add strength for robots that will be used year after year. A RROS-based robot built this way can perform far better than many of the expensive educational robots that are available.

Figure 2.3 also shows how to build a simple "caster" needed on two-wheeled robots. It was made from a short piece of wooden dowel tipped with a chair leg slide as shown in Figure 2.4, but a block of foam board could easily substitute for the dowel. The caster length should ensure your robot's base is level.

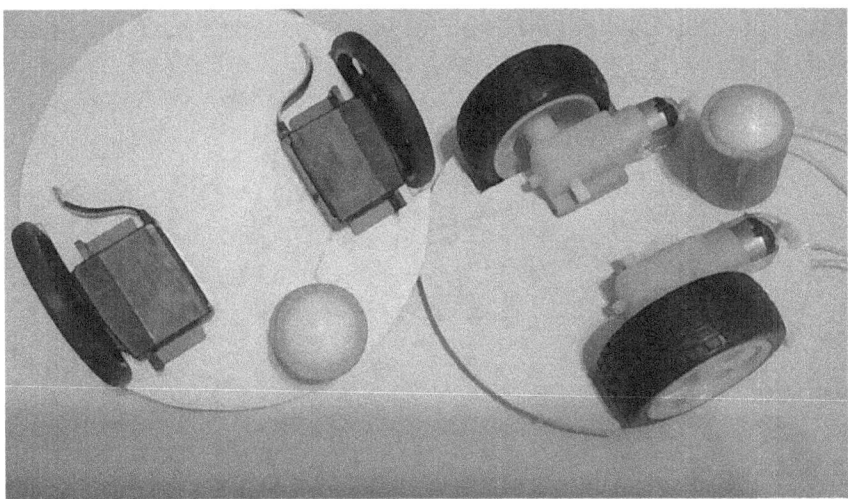

Figure 2.3: The motors can be hot-glued to the foam board.

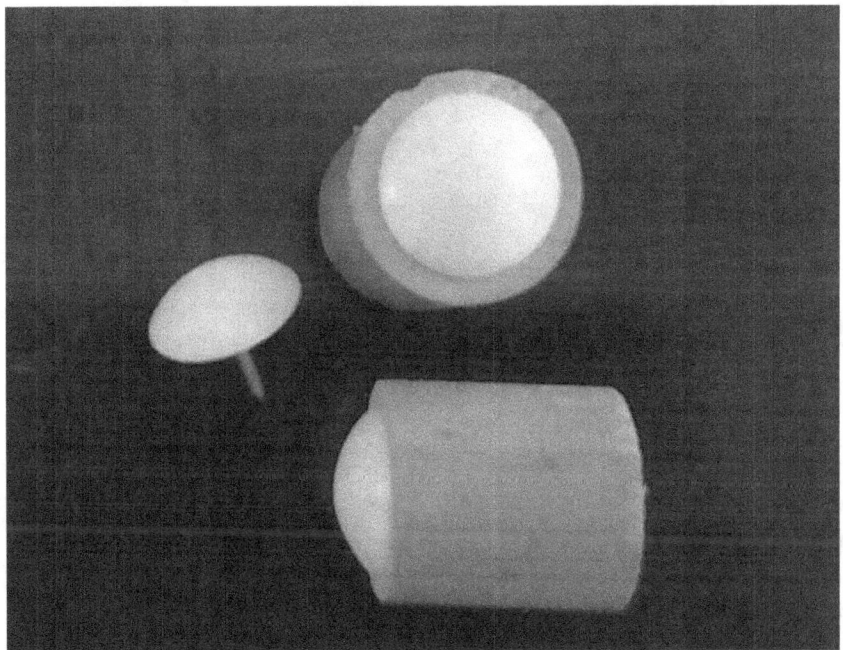

Figure 2.4: A caster can be made from a wooden dowel tipped with a chair leg slide.

As mentioned earlier, just hot-gluing motors to a foam board chassis is easily strong enough for the projects in this book. If you plan on using your robot for more than a year or if it will be subject to rough handling (perhaps younger children) then you might want to add the wood strips mentioned earlier to help support the motors (see Figure 2.3). They are seldom needed though because even if a motor comes loose, a little hot glue can easily solve the problem.

Those readers feeling the need for a much stronger chassis might consider using precut plywood disks available at many craft stores (see Figure 2.5). Of course, you will have to use a saw of some kind to cut out the wheel holes but at least you won't have to cut out the circle itself.

Figure 2.5: Precut plywood disks are available from many craft stores. These are 6.5 inches in diameter just like the foam board bases.

Now that you understand how to construct the chassis, turn to Chapter 3 to get started with the electronics.

Chapter 3
Motor Control Electronics

Once your chassis is constructed, we need to turn our attention to the electronics. In this chapter we will learn how to control your robot's motors.

The schematic shown in Figure 3.7 shows the components necessary to get your robot moving. This schematic is for the DC motor robot. The circuitry for a servomotor robot will be discussed later in the chapter.

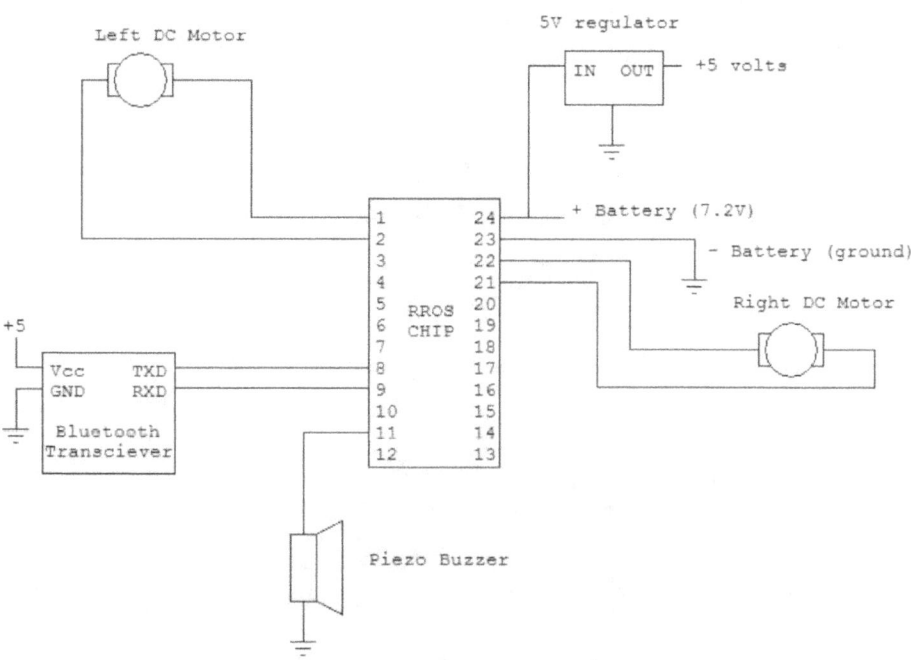

Figure 3.7: This simple circuit will let RobotBASIC control a DC motor powered robot.

The RROS Chip
The primary component in this schematic is the RROS chip (available from RobotBASIC) which is powered by a 7.2 volt battery (more on this later). Notice the two DC motors connect directly to pins 1 and 2 (left motor) and 21 and 22 (right motor). There is also a Bluetooth radio transceiver (available from RobotBASIC and other sources) that allows RobotBASIC to communicate with the RROS chip. The transceiver needs only 5 volts for its power, so a 5 volt, 1 amp regulator (I used a

7805) is used to reduce the main battery voltage down to the correct level for 5 volt components. Having 5 volts available will be handy in future chapters too because the sensors that will be added all need a 5 volt supply. The only other component in Figure 3.7 is a passive piezo buzzer. The robot will work fine without it, but the buzzer will produce beeps at various times to give you feedback when various things happen. Note: Some piezo buzzers are polarity sensitive. If yours is, the + lead should connect to the RROS chip.

Breadboards
The circuit shown in Figure 3.7 is built on a solderless breadboard as shown in Figure 3.8. If you are not familiar with breadboards, read **Appendix A** before continuing here. Even with a solderless breadboard, you will have to solder a few things. The RROS chip, for example comes without the pins soldered to it. Details on soldering the pins can be found in the ***RROS User's Manual*** which can be downloaded from the RROS tab at www.RobotBASIC.org. You may also need to solder pins or wires to the battery cables so they can plug into the breadboard. Take great care not to short the battery pins. Doing so can cause a fire and destroy your battery.

Since the recommended battery is rechargeable, you will need a charger. If it plugs into the original connector on your battery you can simply splice wires to the cable for the breadboard connection. Some chargers come with clip leads that make it easy to attach to the battery.

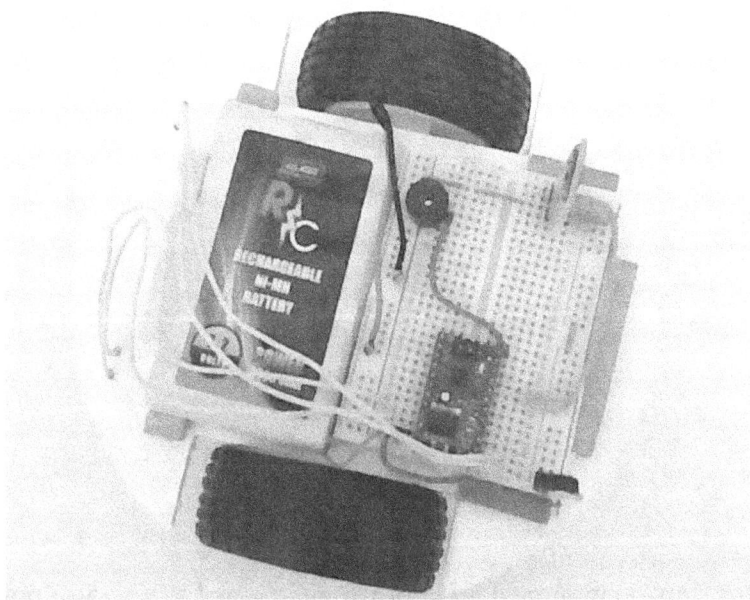

Figure 3.8: A solderless breadboard makes building circuits easy.

Choosing Your Battery
The 7.2V NI-MH battery shown in Figure 3.8 has a capacity of 2000mah. You could use slightly larger or smaller capacities of course, but a battery around this size is light enough for our small robot and will power it for a reasonable amount of time. Notice in Figure 3.8 how small wood strips have been glued around the breadboard and battery to keep them in place.

The circuit in Figure 3.8 sets up the two bus lines near the battery for the battery voltage of 7.2 volts. The 5V output from the regulator is connected to the bus lines on the opposite side of the breadboard. I chose to plug the ground and output pins of the regulator directly into the bus. This means I had to bend up the input wire to the regulator and solder a wire to it (and run the wire to the + line on the 7.2V bus). Alternatively, you could easily insert the regulator elsewhere on the board and run wires from each pin to the appropriate places. Refer to the spec sheet for the regulator you use for its pinout.

Notice how few components are required to build this robot. The two DC motors can connect directly to the RROS chip because it contains the necessary H-bridge circuitry for driving motors as well as the low-level code for controlling them.

The Bluetooth transceiver is in the upper-right corner of Figure 3.8. It allows RobotBASIC to communicate with the RROS chip. Of course, this means your PC must have Bluetooth capability. Most PC's nowadays do, but if you have an older computer, you can add a Bluetooth USB adapter. You will have to pair the PC with the transceiver (more details are provided in the **RROS User's Manual** if you need them). The pairing code for the transceivers sold by RobotBASIC is **1234**. Make note of the port used for Bluetooth communication when you perform the pairing.

Making the Robot Move

With the hardware complete, we can see how easy it is to make the robot move. Look at the program in Figure 3.9. The first two lines of code use an *include* file (provided by RobotBASIC) to set up a lot of constants that make it easier to write RROS programs. Note: This file should be placed in the same directory as the program you are writing. The third line sets a variable equal to the port number of your PC's Bluetooth connection (see the RROS manual if you need help).

The **main** program calls a subroutine to initialize the robot. Once initialized, the robot can be moved using simple commands as shown in the Figure. In this example program, the robot is moved 40 units forward. It then turns $180°$ and moves back to its original position. If either motor moves backwards when tested, just reverse the wires for that motor.

Note: You do not have to have this program's main code offset between the label **main:** and an **end** statement as shown. Doing so though, helps you see where the main code resides. It also puts the main code in the same format used for subroutines (which will be discussed shortly).

```
#include   "RROScommands.bas"
gosub InitCommands    // found in the include file
PortNum = 5 // set to your Bluetooth Port

main:
  gosub InitDCrobot
  rForward 40
  rTurn 180
  rForward 40
end
```

Figure 3.9: This simple program shows how to control the robot's drive wheels.

If the simulator was initialized (instead of the DC motor robot) the units for **rForward** are screen pixels. The simulated robot is 40 pixels in diameter so the command **rForward 40** will move the simulated robot a distance equal to its diameter. Ideally, the same command should move the real robot a distance equal to its diameter.

Precise control of a robot's movements normally requires the drive motors to be equipped with wheel encoders that produce pulses as the wheels turn. The robots computer (the RROS chip in this case) can count these pulses to monitor the robot's movements. Even though the RROS chip supports wheel encoders, but there is no reason to add the cost or complexity to a beginner-bot because the RROS chip can be calibrated for timed responses when encoders are not available. The robot's movements will be less accurate without encoders but, as you will see, the robot can still function in an acceptable manner, especially when other sensors are used.

Initializing the DC Motor Robot
Look at the initialization subroutine in Figure 3.10. The first line in the subroutine sets up the communication link with the Bluetooth port assigned earlier. The second line initializes the RROS chip (if the Bluetooth link has been enabled with **rCommPort**). Next, we need to tell the chip we are using small DC motors and set the speed we want the robot to use (0-100). In general, you will want a slow speed for projects in this book.

Since our robot is not using wheel encoders it will almost certainly drift to the right or left when moving forward because no two motors are exactly alike. The RROS allows you to reduce the speed for either the left or right wheel, as shown to correct for drift. For my robot, I needed to reduce the speed of the right wheel by 5%. You will need to experiment to find what works for your robot.

Subroutines are executed by calling them with a **gosub** statement (see Figure 3.9). When the **return** statement is encountered, execution resumes at the line following the calling **gosub** statement. The **InitDCrobot** subroutine should be typed into the RobotBASIC editor following the code of Figure 3.9.

```
InitDCrobot:
  rCommport PortNum
  rLocate 10,10
  rCommand(MotorSetup,SMALLDC)
  rCommand(SetSpeed,17)
  rCommand(SetReducForwRight,5)
  rCommand(SetMoveTime,34)
  rCommand(SetRotationTime,33)
return
```
Figure 3.10: The RROS chip must be properly initialized to work accurately.

The next two lines in Figure 3.10 establish how long the motors will run when the robot is moving forward and making turns. As mentioned earlier, if you issue an **rForward 40** command, the robot should move a distance equal to its diameter. If it is moving too short or too far, adjust the **MoveTime** parameter for the best results. The robot should also turn 90° when executing an **rTurn 90** command. Increase or decrease the rotation parameter to produce accurate turns.

It is important to realize that timed movements and drifting corrections will never be as accurate as wheel encoders (which you can add later, if desired, as your skill-level and budget increases). And even when you think you have the timing calibrated, a drop in the battery voltage can screw everything up again. But remember, this is an entry-level robot so you can't expect perfect performance. The good news is that once we add sensory capabilities the robot can be programmed to use sensor data to constantly correct its actions. Such corrections can produce a great performance for many behaviors even without wheel encoders.

A Servomotor Robot
If you prefer servomotors over DC motors, the RROS can accommodate either. Everything in Figure 3.7 is the same except for the motor connections. Just interface the servomotors as shown in Figure 3.11. Servos are connected with three wires. Ground is usually black or brown and the positive supply voltage is usually red or orange. The control signal is supplied over the third wire which is often white or yellow.

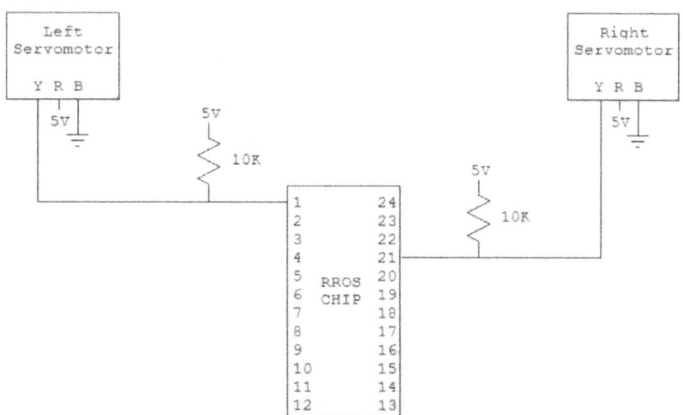

Figure 3.11: The motor connections for servomotors are the only changes needed to Figure 3.7.

Controlling the Servomotor Robot
The same programs used to control the DC robot can control the servomotor version as long you initialize the RROS chip properly. Figure 3.12 shows an appropriate subroutine for the servomotor initialization. Drift and movement times are set just like a DC motor. Servomotors though, require one additional setup.

Normally, a continuous rotation servo will stop moving when it receives a pulse whose width is in the center of its range. If either of your robot's servomotors continue to rotate when the robot should be stationary, you can correct it by increasing or decreasing the value of its **StopOffset** from its default value of 128 (see Figure 3.12).

```
InitServoRobot:
  rCommport PortNum
  rLocate 10,10
  rCommand(MotorSetup,SERVOMOTORS)
  rCommand(SetLeftStopOffset,168)
  rCommtand(SetSpeed,15)
  rCommand(SetMoveTime,26)
  rCommand(SetRotationTime,30)
return
```

Figure 3.12: This routine initializes the RROS chip for servomotors.

Now that you have the main circuitry built for your robot, turn to Chapter 4 to start your first project.

Chapter 4
Controlling the Robot's Movements

Now that we have a robot with drive motors, let's create a project that does something a little more exciting. The program in Figure 4.1 will make RobotBASIC's simulated robot move in various shape patterns like triangles, squares, octagons, etc. Just set the variable **NumSides** equal to the number of sides you want your shape to have. If you want a triangle, for example, set the number of sides to 3.

    The second line in the program prints the characters inside the quotes exactly as written followed by the value of **NumSides**. As you will see, this is a nice touch because it lets the output screen show the user the number of sides being drawn. Next, the variable **TurnAngle** is set to 360/**NumSides**. This value will be used to control how much the robot should turn at each of the corners of the polygon it is drawing. The robot cannot turn less than one degree, so some shapes (11 sides, for example) will not be perfect due to rounding errors.

```
NumSides = 9
print "Number of sides = ",NumSides
TurnAngle = 360/NumSides
rLocate 100,300
rInvisible GREEN
rPen DOWN
for i=1 to NumSides
  rForward 100
  rTurn TurnAngle
next
print "All Done"
```
Figure 4.1: This program lets the simulated robot draw shapes.

Once the robot is located on the screen, the pen is dropped, but there is also a new command that can specify one or more invisible colors. The first invisible color (GREEN in this case) is automatically used for the pen. Making this color *invisible* is essential in this program because the robot returns to its original position during its movement and this would cause the robot to travel over the line it has previously drawn. If the line is not invisible, the robot will see it as an object on the screen and colliding with it will cause an error. Future chapters will discuss this idea in far more detail and utilize this feature to create exciting projects.

Loops in Programming

The real work of moving the robot is handled by a **FOR-NEXT** loop. As we will see in future chapters, most programs are full of loops that perform a series of actions over and over. In this case, the **FOR** statement marks the beginning of the loop and the **NEXT** marks the end of it. The **FOR** starts by setting the variable specified (**i** in this case) equal to 1. It then executes the statements that follow the **FOR** until it gets to the **NEXT**. At that point, the variable initialized by the **FOR** will be incremented. If the new value is greater than **NumSides** (see the **FOR** statement) the program will terminate the loop and continue execution with the line following the **NEXT** (or terminate the entire program if **NEXT** is the last line in the program. In this case, the last thing the program does is print the phrase **All Done**. Notice the phrase (often called a string in programming, as in a string of characters) is in quotes to prevent RobotBASIC from thinking it is a variable name or a command.

If you set the number of sides to 6 and run this program, you will see the output shown in Figure 4.2. Type in the program on your computer and run the program to watch the robot actually draw the shapes specified by **NumSides**.

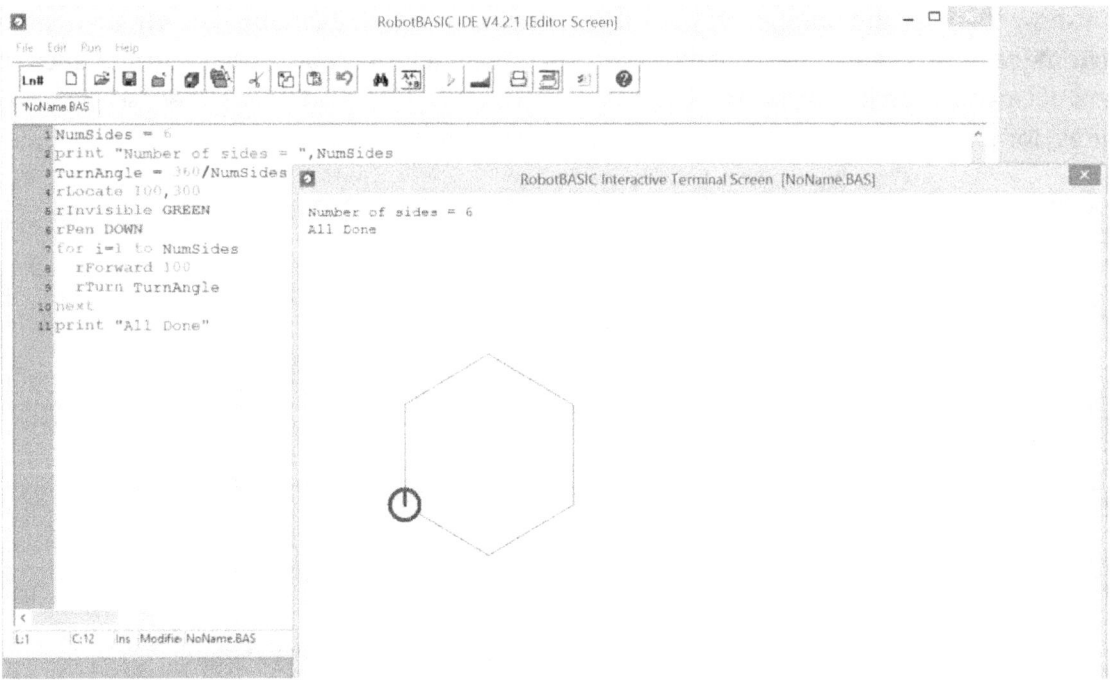

Figure 4.2: The simulated robot can draw.

If you try to create shapes with 14 or more sides, the shape will be so large that the robot will collide with a wall causing an error (the edge of the screen is consider to be a wall around the robot's environment). This can easily be fixed by reducing the distance traveled by the robot on each pass through the loop. For example, change **rForward 100** to **rForward 50**.

Modifying the Program

Now that you know a little about RobotBASIC, it is time for you to start experimenting with programming – especially if you have never programmed before. In addition to modifying the value of **NumSides**, try reducing the distance the robot travels or changing the invisible color. You can use things like **RED, BLUE, YELLOW, CYAN, BROWN, and GRAY**. RobotBASIC can actually produce thousands of colors, but this is enough for now. Try all of the above ideas and examine the shapes drawn. Study your modified programs to make sure you see why your changes produced the results you see.

Try to modify the program so it draws a circle. **Hint**: Reduce the length of travel and increasing the number of sides. You might even try to locate the robot so that it starts at a different place on the screen. Don't worry if your program produces errors or if you start having a lot of questions. Your experimenting cannot damage RobotBASIC. The worst thing that might happen is that you'll get excited about programming and become impatient to explore future chapters.

The Simulator vs a Real Robot

You may well be wondering why all this discussion on the simulator because you bought this book to learn how to control a real robot. Let's modify the program in Figure 4.1 so it can make your REAL robot move in a multi-sided pattern, just like the simulator. Using a simulator is great for hobbyists because it allows you to get your program working without having a real robot bumping into walls or raging out of control. We will allow the program to control a real robot only after it successfully controls the simulation.

Using a simulator also a great way for a school to teach robotics without spending a lot of money. Every student can program their own simulated robot and then when someone gets their program working, they can use their program to control a single real robot.

The program in Figure 4.3 will control *either* the simulation or a DC motor powered robot depending on the value of the variable **Real**. If the value is **False**, the simulator will draw a polygon with the specified number of sides just like Figure 4.2. If the value is **True**, the real robot will move in a similar manner. As mentioned earlier, our beginner-bots do not have wheel encoders so they will not move as accurately as the simulator. More on this shortly. Right now, let's look at the code in Figure 4.3 to see how it can control either robot.

The first three lines in the program are the same as Figure 3.9 back in Chapter 3. The next line sets up the variable **Real** to be **True** or **False** depending on whether we want the program to control a real robot or the simulation. Most of the main program is the same, except where the robot is initialized. In this program, the initialization is handled by an **IF-ELSE-ENDIF** control structure (*construct* for short).

RobotBASIC has several types of **IF** statements. In this case, if the *expression* following the **IF** is true, the statements between the **IF** and the **ELSE** are executed. If the expression is false, the statements between the **ELSE** and the **ENDIF** are executed. In either case, execution continues with the line following the **ENDIF**.

Look at the **IF** section of code in Figure 4.3. If the variable **Real** is equal to **False**, then the simulator is initialized just as it has been in previous examples. If **Real** is **True**, a **gosub** statement calls a subroutine to initialize our real robot with DC motors, again, just as we have in previous examples. If you are using a servomotor powered robot, simply call the subroutine of Figure 3.12 from Chapter 3 instead of initializing the DC robot. You will, of course, have to type in that subroutine instead of the one for the DC motor initialization. Remember, your initialization should set up the drift and move/rotation parameters appropriate for YOUR robot, not mine.

```
#include  "RROScommands.bas"
gosub InitCommands    // found in the include file
PortNum = 5 // set to your Bluetooth Port
Real = False // set to True to control the real robot

main:
  NumSides = 9
  print "Number of sides = ",NumSides
  TurnAngle = 360/NumSides
  if Real=False
    rLocate 100,300
    rInvisible GREEN
    rPen DOWN
  else
    gosub InitDCrobot
  endif
  for i=1 to NumSides
    rForward 100
    rTurn TurnAngle
  next
  print "All Done"
end

InitDCrobot:
  rCommport PortNum
  rLocate 10,10
  rCommand(MotorSetup,SMALLDC)
  rCommand(SetSpeed,17)
  rCommand(SetReducForwRight,5)
  rCommand(SetMoveTime,34)
  rCommand(SetRotationTime,33)
return
```

Figure 4.3: This program will cause either the simulator
or a real robot to move in a polygon shape.

Look carefully at the format of the program in Figure 4.3. Notice how the code for controlling the robot has been organized into two modules (**main** and **InitDCrobot**) and that the code is indented to help identify each section. Notice also that the code inside the **FOR-NEXT** construct as well as the two sections of the **IF-ELSE-ENDIF** are similarly indented. This type of structure helps you see where the code inside loops and decisions begins and ends. Such organization is NOT required, but it does make programs far easier to read and understand – not just for you but for anyone else reading your code. Companies that hire programmers generally require them to follow company-specified formats so that all programs within the company have the same look and feel. This practice has proven to make programmers more productive and the whole process more efficient.

Change the value of the variable **Real** appropriately and run the program of Figure 4.3 to verify that it will control both the simulation and your robot. Modify the program so that the robot moves to create shapes of different sizes and number of sides. Remember, the real robot's movements will not be accurate, but it will move in a manner similar to the simulation.

In Chapter 5, we will begin adding sensors to our robots. This will allow them (with proper programming) to perform many actions accurately even without wheel encoders. Since we are going to develop our programs on the simulator though, it can be nice to have it respond similarly to a real robot. Actually, this is easy to do.

Add the following line to the end of the code that initializes the simulated robot in Figure 4.3.

```
rSlip 10
```

This command tells the simulation to add a random error of up to 10%. This will affect the distance traveled as well as the accuracy of turns. The error is referred to slip because the wheels on a real robot might slip on a waxed floor, causing a similar error. If you run the program with slip added, the output will look similar to Figure 4.4. The error is random, so every time you run the program it will be different.

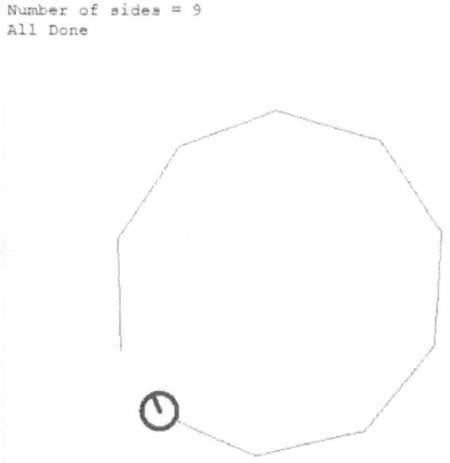

Figure 4.4: Adding slip to the simulation makes its movements more realistic.

In future chapters, we will be able to use the slip feature to make the simulation more realistic. This will allow us to develop programs that utilize sensory information to constantly correct the robot's movements. Turn to Chapter 5 and let's start learning about sensors.

Chapter 5
Adding a Ranging Sensor

The robot's you have seen so far in this book are not much better than toys. A *real* robot should have the ability to interact with its environment using sensor information. How well it interacts is a reflection of its *intelligence*. As we proceed through this book, we will add more sensors to our robot and learn how to create programs that allow it to utilize sensor data to accomplish goals such as avoiding objects and following a line or wall.

Adding a Ranging Sensor
Let's start by adding an ultrasonic ranging sensor to the DC robot discussed in previous chapters. Figure 5.1 shows Parallax Ping sensor. It can be purchased from www.Parallax.com. The sensor works by sending out high-frequency sound waves (above the human hearing range) and measuring the time it takes for the sound to bounce off objects and return to the sensor.

Figure 5.1: This ultrasonic sensor can measure distances to objects.

The Ping sensor has three pins for connecting it to our robot's electronics. One pin is labeled GND for ground, one is 5V for 5 volts and the last pin is SG and provides the data *signal* that will connect to the RROS chip. Normally this sensor can be complicated to use by novices, but the RROS chip will do all the work for you.

Other than adding the Ping sensor to our robot, no physical modifications need to be made. The Ping-equipped robot is shown in Figure 5.2. The pins on the sensor are just plugged into the breadboard using three unused rows of holes. The silver, round cylinders (a speaker and microphone) should face away from the front of the robot, as shown.

Chapter 5: Adding a Ranging Sensor

Figure 5.2: The Ping sensor can be mounted on your robot as shown.

Figure 5.3 shows the updated schematic from Chapter 4 with the Ping sensor added. Only three connections are required (5V, ground, and signal). Just wire the Ping pins to the appropriate placed on the breadboard. These Ping connections are used for servomotor powered robots too. Think about how easy it is to add a Ping to your robot – you just plug in the sensor and three wires. As we add additional sensors to your robot, you will be more and more impressed with how easy to RROS system is to use.

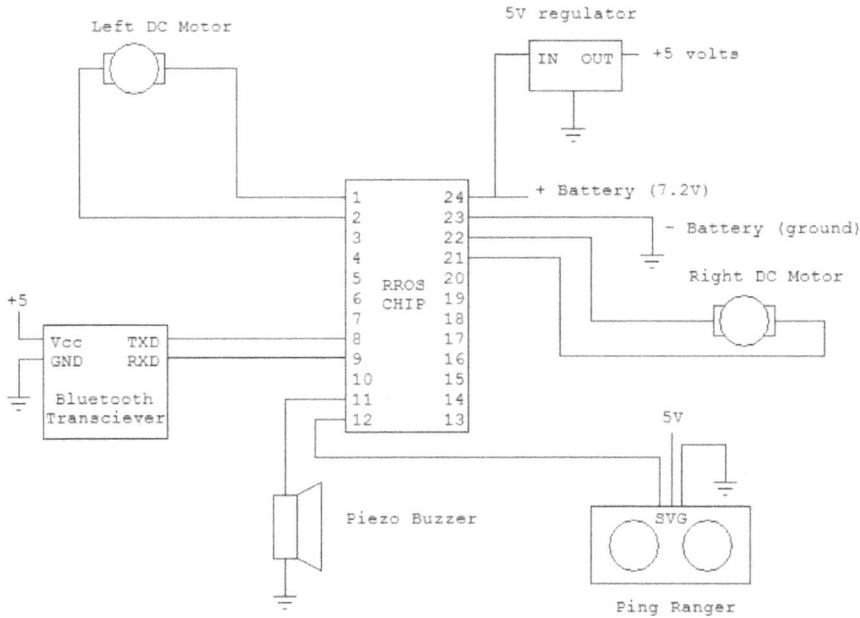

Figure 5.3: Wire the Ping sensor to your robot as shown above.

Obtaining and Using the Ranging Data

The program in Figure 5.4 moves the robot forward until it is 5 inches from an obstacle (the units returned by **rRange()** are ½ inch for the real robot). After initialization, a **while**-loop continually moves the robot forward in tiny increments while the expression in the **while**-statement is true. This means that all the statements between the **while** and the **wend** will be executed as long as the distance read by the sensor is greater than 10.

```
#include   "RROScommands.bas"
gosub InitCommands
PortNum = 5 // set to your Bluetooth Port
Real = False // set to True to control the real robot

main:
  if Real=false
    rlocate 400,300
    rInvisible GREEN // remove these 3 lines for no trail
    rPen DOWN
  else
    gosub InitDCrobot
  endif
  while rRange()>10
    rForward 1
  wend
end

InitDCrobot:
  rCommport PortNum
  rLocate 10,10
  rCommand(MotorSetup, SMALLDC)
  rCommand(SetSpeed,17)
  rCommand(SetReducForwRight,5)
  rCommand(SetMoveTime,34)
  rCommand(SetRotationTime,33)
  rCommand (SensorSetup, PING)
return
```

Figure 5.4: This program will moved the robot forward until it reaches a wall.

Real vs Simulation

The program in Figure 5.4 will work with both the real robot and the simulator. Let's start by setting **Real** to **False** and watch the simulated robot. It will move forward until it is 10 pixels from the upper wall on the computer screen, as shown in Figure 5.5. If you set **Real** to **True**, the real robot will move forward until it is 5 inches (ten ½ inch units) from some object blocking its path.

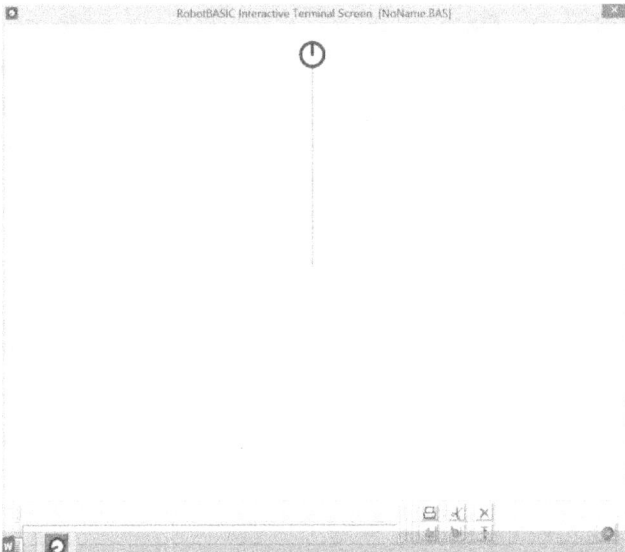

Figure 5.5: The robot was initialized in the center of the screen and moves until it reaches a wall.

If you are using the real robot place objects in front of it to verify it always stops when it sees them. Note: The sound waves must bounce back to be seen, so it is possible that a soft stuffed animal might absorb the sound, or an angled wall might reflect it away from the robot.

If you are using the simulator, you can actually draw objects to block the robot's movements. Enter the following line right *before* the **rLocate** statement in Figure 5.4. When you run the program you will see the screen in Figure 5.6. Note: The **rectangle** command draws a rectangle defined by its upper-left (300,50) and lower-right (500,100) corner coordinates (see the RobotBASIC help for more information).

```
rectangle 300,50,500,100
```

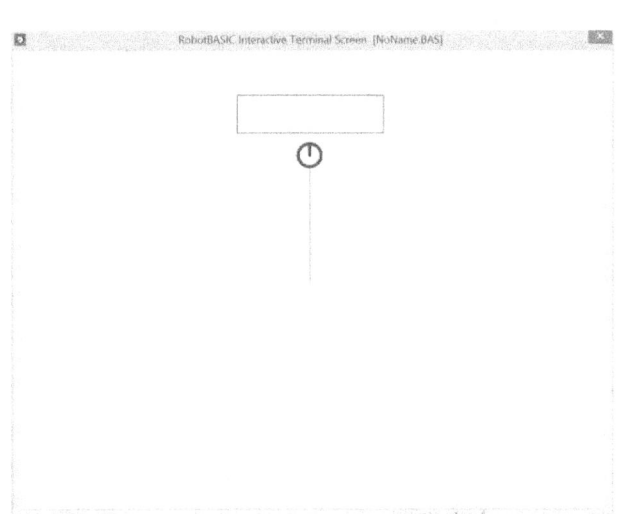

Figure 5.6: The simulated robot will stop when it reaches any object on the screen.

As you can see, the robot (both real and simulated) will stop when it encounters an object. Try moving the object on the screen or locating the robot in a lower position to test the simulation. Make the object red by executing the statement **SetColor RED** before drawing the rectangle. When you see that the robot still stops, make the color **GREEN** and watch what happens. Can you explain why?

When you feel comfortable with everything in this Chapter, turn to Chapter 6 to learn more about using **rRange** to control your robot's movements.

Chapter 6
Avoiding Objects While Roaming

In previous chapters we have seen how to construct a robot chassis and the electronics for powering the motors and interfacing with a Ping ranging sensor from Parallax. You have also learned a little about programming with RobotBASIC. In this chapter, we are going to get serious about programming and really start to learn what it is all about.

The Goal
Our final goal for this chapter will be to program our robot to roam around a room, avoiding objects in its way without any human help. The basic idea is simple. The robot will use the Ping sensor to determine if there are any objects in its path. If not, it will move forward a short distance and test again for objects. This sequence will be repeated over and over, making the robot continually move forward if nothing is in its way.

If an object is detected during this sequence though, the robot will turn away from the object and start the whole process over again. As when developing any program, we will start with a basic strategy and improve on it as we see how the robot reacts to our plan.

Using the Simulator for Development
We will use RobotBASIC's simulated robot to design and test our program. Once we get it working properly, we will modify the program so it can control the real robot. The program in Figure 6.1 is our starting point. Notice the main program is very short because it simply calls three other modules to handle the tasks that needed to be done. One of these tasks creates our simulated room (the environment where our robot will roam). Another module initializes the robot (for now, just the simulation). Finally, the **Roam** module will perform the work needed to accomplish our desired goal.

The **CreateRoom** module draws circles and rectangles that serve to block the robot's movement. Circles use coordinates just like rectangles. A circle (or ellipse) is drawn inside the rectangle defined by the specified coordinates. Notice these commands now include colors for the border and interior of the shapes. Many of the commands discussed in this book have additional features like this so it can be valuable to read through RobotBASIC's help file when you have time. It is hundreds of pages, so don't expect to do it quickly. A printed version is available on Amazon.com for those that would prefer the convenience of a standard book.

The **InitRobot** module initializes the simulation and locates the robot in the center of the screen. It also drops the robot's pen so it will leave a trail to make its movements easy to see.

Comment out the pen-related lines if you don't want a trail. If you place two slashes (**//**) at the beginning of a line, the rest of the line is ignored.

```
Main:
  gosub CreateRoom
  gosub InitRobot
  gosub Roam
end

CreateRoom:
  LineWidth 3
  Rectangle 200,0,300,250,RED,BLACK
  Rectangle 550,350,799,475,RED,BLACK
  Circle 100,350,300,450,RED,BLACK
  Circle 500,150,650,200,RED,BLACK
  Circle 350,550,450,650,RED,BLACK
return

InitRobot:
  rLocate 400,300
  rInvisible GREEN
  rPen DOWN
return

Roam:
  // left blank to start
return

CheckForObjects:
  NoObjects = TRUE
  for angle = -15 to 15
    if rRange(angle)<100 then NoObjects = FALSE
  next
return
```

Figure 6.1: This program is the starting point for our roaming robot.

The Roam module is currently blank (see the comment in the program). We will develop it shortly. There is also a **CheckForObjects** module, but that too will be addressed soon.

The Robot's Environment
If you run the program, you will see the screen shown in Figure 6.2. Notice the five shapes that have been placed throughout the screen. Examine the code in the **CreateRoom** module of Figure 6.1 to see how each of the shapes are created. Notice also that the robot is located in the center

of the screen as expected. All we need to do is create the Roam module and our program will be done.

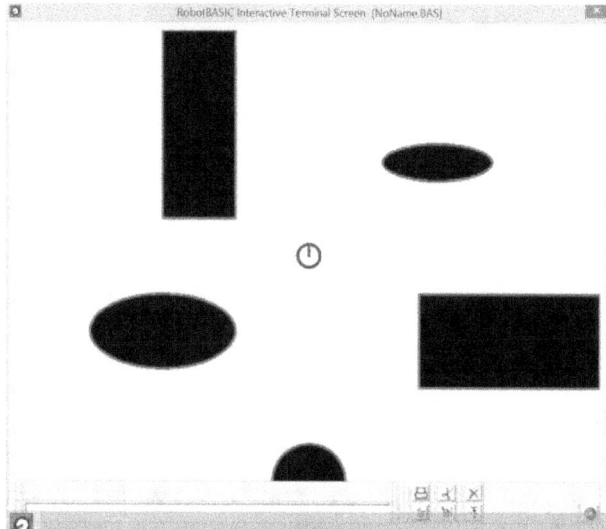

Figure 6.2: This is the environment for the simulated robot.

Remember, our goal is for this program to work with the real robot. In order for that happen seamlessly, both the simulation and the real robot must respond similarly to RobotBASIC programming statements. For that reason, we need to address an issue with the Ping sensor.

Ultrasonic vs Infrared
As discussed in previous chapters, the Ping sensor works by sending out sound waves and watching for them to bounce off objects and return. Ping is not the only type of ranging sensor available. Some rangers use a beam of infrared light to detect objects and measure how far away they are. The RROS chip actually allows you to use a variety of infrared (IR) and ultrasonic sensors made by different companies. To keep things simple, this book will only utilize the Ping for ranging, but you can refer to the *RROS User's Manual* when you want to expand beyond the scope of this text.

The reason this is important is that while both types of ranging sensor measure distance, the beam on the IR sensor is very narrow and can only detect objects directly in front of the sensor. The sound waves on ultrasonic sensors though, spread out in a cone shape. Both sensors can only detect objects hit by their beam, but the wider cone-shaped ultrasonic beam allows it to detect objects that are off to the side slightly as shown in Figure 6.3. In many situations, this makes ultrasonic rangers more useful.

Unfortunately, the simulated robot handles its ranging operation more like an IR sensor (it only checks for objects in a straight line). Fortunately though, it is possible to tell the **rRange()** function to look slightly left or right by giving it an angle in the parenthesis following the command itself.

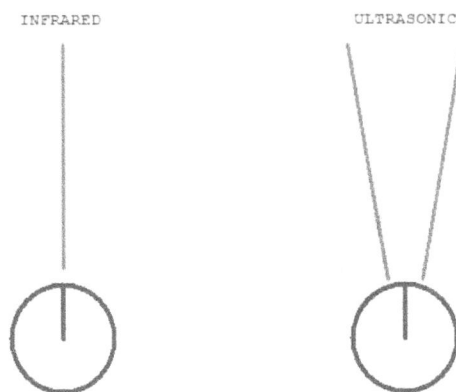

Figure 6.3: Ultrasonic rangers have a wider beam than IR.

Emulating an Ultrasonic Ranger

The **CheckForObjects** module in Figure 6.1 makes it possible for the simulator to appear to have an ultrasonic ranger. Let's look at the module to see how it works. It starts by assuming there are no objects to be detected and conveys that information by setting the variable **NoObjects** to **TRUE**. It then uses a **FOR**-loop to make the variable **angle** assume values from -15 to +15. Inside that loop, the **rRange(angle)** function looks at each angle within a 30° cone (-15, -14, ... 15) to see if it sees an object closer than 100 pixels at any of those directions. If it does, it sets the variable **NoObjects** to **FALSE** to indicate that an object was detected. Notice this is done with a new form of the **IF** called the **IF-THEN**.

The **IF-ELSE-ENDIF** is nice if you have one or more statements that need to be executed when the parameter for the **IF** is *either* true or false. The **IF-THEN** is a shorter, simplified version that makes it easy to perform one statement if the **IF** parameter is true. For **FALSE** conditions, the statement following the **THEN** is just ignored. RobotBASIC has several types of IFs and loops (read the help file for more information).

Every time the **CheckForObjects** module is called, it will update the value of **NoObjects**. Other portions of the program can then use an **IF** to see if any objects were detected in a 30° cone just like an ultrasonic ranger might have. Note: The cone for a real ultrasonic sensor will not be exactly 30°, but this approximations should work fine.

Building the Roam Module

We are now ready to build the Roam module needed for Figure 6.1. Remember, it should continually move the robot forward unless it sees an object. When objects are detected, the robot should turn away. The module shown in Figure 6.4 avoids the objects in its way by turning the robot to the left by 10°.

The **while**-loop inside the **Roam** module runs forever (unless you terminate the program), because its argument is always true. Inside the loop, the first thing that happens is the **CheckForObects** module is called to check if anything is seen. This module, as you recall, sets the value of **NoObjects** to **TRUE** or **FALSE** to indicate if objects have been detected.

```
Roam:
  while TRUE
    gosub CheckForObjects
    if NoObjects
      rForward 40
    else
      rTurn -10
    endif
  wend
return
```

Figure 6.4: This version of **Roam** will always turn the robot left to avoid objects.

If there are no objects, the robot moves forward 40 pixels, a distance equal to its diameter. This distance is reasonable, because the **CheckForObjects** module looks for obstacles a distance of 100 pixels which is 2 ½ times the diameter of the robot. This should give the robot enough time to make an avoidance maneuver.

If you run the program you will see the robot *starts* roaming as shown in Figure 6.6. Since the Figure only shows the initial movements of the robot, it is easy to see that it always makes left turns when objects are encountered If you don't stop the program (as I did to get this Figure) the robot will continue roaming around the screen avoiding objects by making left turns.

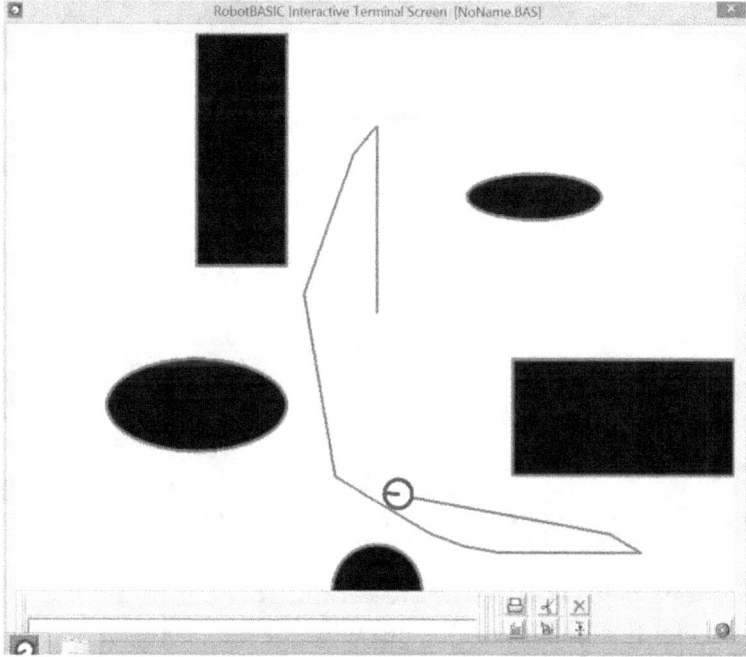

Figure 6.6: The robot avoids objects by always turning left.

Another thing that is worth noting about this program is that the robot's path is the same every time the program is run. That is true because the robot is reacting to its environment and the environment is always the same. You could add slip, as discussed in Chapter 4, but since the robot is constantly reacting to obstacles, the path will not change much.

Improving the Roaming
What we really need is for the robot to do is sometimes turn left and sometimes turn right. We could make that decision many ways. For example, we could let the robot turn both left and right and take range measurements, and then turn toward the direction with the longest range (meaning the other direction had an obstacle closer). This is really not that hard to do, but it could be a little confusing for a novice programmer. For that reason, let's try a different approach.

Let's modify the **Roam** module so that it chooses its turn direction randomly. Figure 6.7 shows one way to do that. Since the **ELSE** section of the **IF**-statement is executed when the robot turns, this is the only part of the code that needs to be modified. In this case, another **IF-ELSE-ENDIF** is *nested* inside the **ELSE** portion of the first **IF**. Both **IF** and loop control structures can be nested inside each other as many times as you want. Notice that the new **IF** is indented again to make it easier to identify. In fact, look carefully at the indenting in Figure 6.7. The entire contents of the **Roam** module is indented to help you see where it begins and ends. Then, the entire contents of the **WHILE**-loop is indented again. Inside the **WHILE**, the two sections of the first **IF** are indented even more and then even more for the section of the second **IF**.

```
Roam:
   while TRUE
      gosub CheckForObjects
      if NoObjects
         rForward 40
      else
         if random(100)>50
            rTurn 10
         else
            rTurn -10
         endif
      endif
   wend
return
```

Figure 6.7: This version of Roam randomly turns both left and right.

If you don't think indenting helps you see where various sections of the program begin and end, look at Figure 6.8. It is identical to Figure 6.7, but without indenting. Imagine how difficult it would be to read programs that were written this way. It would even be difficult to find a module you were looking for.

Let's get back to understanding how Figure 6.7 makes the robot turn both left and right. It uses a new statement called **random()** that, in this case, generates a random number between 0 and 100. If that value is greater than 50 (which it should be about half the time if the numbers are truly random) then the robot turns right. Otherwise, it turns left.

Unfortunately, even though this sounds like an easy fix, it does not work very well at all. Before reading further, see if you can think of any reason it should not work. Problems like this one are not uncommon in programming, especially for beginning programmers. The more experience you get, the better you will be able to predict how changes you make will affect a program's operation.

```
Roam:
while TRUE
gosub CheckForObjects
if NoObjects
rForward 40
else
if random(100)>50
rTurn 10
else
rTurn -10
endif
endif
wend
return
```

Figure 6.8: Indenting really helps identify sections of a program.

Problems with the Program

If you run the program with the **Roam** module shown in Figure 6.7, you will see that the robot tends to stop a lot and just vibrate between left and right turns. Eventually it moves on, but the constant vibration behavior is annoying and takes away from the idea that the robot might be even a little intelligent. Try it on your computer to see it for yourself. Can you determine why this behavior is so different from the original program?

The answer is not immediately apparent to most people new to programming. In the original program, if the robot turned left and the object was still in the way, then it just turned left again – and continued to do so until the area was clear. Look back at Figure 6.4 and examine the code carefully to see how the looping causes the robot to constantly turn left until it no longer sees an obstacle.

The version in Figure 6.7 works differently. When an object is seen, the robot might turn either left or right but if the pathway is still blocked, it makes another turn, but there is a 50% chance that it will just turn back to its original blocked position. Obviously since the turns are random, the robot might get lucky and immediately avoid an object in its path. But, it is also likely that the robot will end up randomly turning left and right numerous times before it eventually chooses the right random turns to escape a blocked situation.

Fixing the Problem

Can you think of how this problem might be solved? Spend a few minutes thinking about it before moving on. If you are with a group of people (perhaps in a classroom) work together to see if you can find a solution. Don't think about how to *program* a fix. Instead, think of the robot's behavior and what it might do differently to solve the problem.

As with all programming problems, there are many possible solutions. The one I chose to implement was to let the robot make left turns *for a while*, then make right turns *for a while*. This is very different than making the decision *every time* a turn need to be made.

Think of it like this. A left-handed person might favor making left turns and a right-handed person might prefer right turns. My solution for our robot is to let it be left-handed for a while and then right-handed for a while. Furthermore, I think the amount of time the robot is either left or right-handed should be random. This is important because it means the robot will behave differently every time we run the program and that seems preferable for this type of program.

Coding vs Developing Algorithms

Notice that the discussion above did not talk about *how* to program the robot to do anything. Instead, it simply expressed an idea of *what* the robot should do. In general, programmers should decide *what* has to be done first, then decide *how* to write the code to do it. As you become more and more proficient in a computer language the difference between the above two ideas will blur, but in the beginning, you should certainly see them as distinct processes.

The plan for what needs to happen is called an *algorithm*. Writing a program to implement the plan is called *coding*. Programming is really the process of creating an algorithm and implementing that using the code required for the language you are using. If you learn how to create algorithms, then you can program with many language by just studying their instructions.

Our next step is to see how we can make the robot change between right and left turns for a random amount of time. Look at Figure 6.9. The actual **rTurn** statement uses **dir*10** to control the turn itself. If the value of **dir** is **+1** the robot will turn right. A value of **-1** will turn the robot left. At the beginning of the module, the variable **dir** is set equal to **1**. Inside the loop, a random statement multiplies **dir** by **-1** about 5% of the time.

If **dir** is 1, multiplying it by -1 will make it a -1. If it is already -1, multiplying it by -1 will make it positive again. This means the robot will change the style of its turns about 5% of the time and that should eliminate the vibrating movements seen in the last module used. Make these modification to your **Roam** module and verify the program now works as expected.

This program is not necessarily perfect, but it does seem to perform fairly well based on our original goals. It certainly is possible though that the robot, if it runs long enough, might eventually bump into an object for some unusual position of the robot in the room. And since our robot moves with some randomness, it should eventually find itself in every possible situation (given enough time).

```
Roam:
  dir = 1
  while TRUE
    if random(100)<5 then dir = dir*-1
    gosub CheckForObjects
    if NoObjects
      rForward 40
    else
      rTurn dir*10
    endif
  wend
return
```

Figure 6.9: This version of **Roam** works well.

Let your robot run awhile, eventually you should see something like Figure 6.10. The robot roamed perfectly for a reasonable period of time, but then it just nicked the corner of one of the obstacles causing the error.

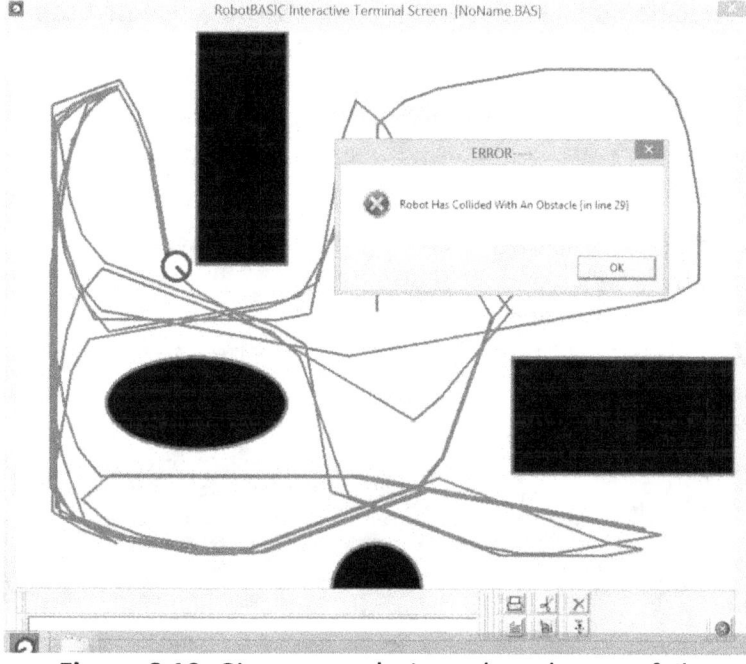

Figure 6.10: Given enough time, the robot can fail.

Improving the Robot

Can you think of how to solve this problem? Would it help to make the robot turn 15° instead of 10? Would 20 be better? Would it crash quicker if it only turned 5°? What if you made it move a shorter distance before checking the sensor again. Right now it moves 40 pixels. Would 20 be better or perhaps 10? What if it could see objects further away than 100 pixels? Would increasing the scan distance from 100 to 125 be enough to solve the problem?

Actually, some of the suggested modifications above will help a lot, others, not so much. Try making various modifications to see if you can make the robot roam properly without bumping into objects. After trying each type of modification alone, see if doing a little bit of each of them together is better or worse. If the robot tends to get stuck doing the same thing after some of your modifications, consider changing how often it converts between left and right turns. When you have your program working to your satisfaction, come back here to see how we can use it to control a real robot.

Making the Real Robot Roam

Figure 6.11 shows how to modify the program so it will control the real robot. A few lines have to be added before **Main** just as before. Inside of the main section, an **IF-THEN** ensures that that the room is only created on screen when using the simulation. The **InitRobot** module now initializes the proper robot. Finally, the **CheckForObjects** module handles the real robot differently. For one thing, it does not need to take a bunch of reading for the real robot because the Ping sensor already covers a cone-shaped area. Also, we want the Ping to look out a distance equal to about 2 ½ times the real robot's diameter (just like the simulation) so we use a parameter of 26 which is 13 inches. Test the program with your robot and make modifications like you did for the simulation if it collides with objects. Notice also that the **InitDCrobot** from previous programs has been included for clarity. You should use the initialization module you have prepared for your robot.

```
#include  "RROScommands.bas"
gosub InitCommands
PortNum = 5 // set to your Bluetooth Port
Real = False // set to True to control the real robot

Main:
  if Real=False then gosub CreateRoom
  gosub InitRobot
  gosub Roam
end

CreateRoom:
  LineWidth 2
  Rectangle 200,0,300,250,RED,BLACK
  Rectangle 550,350,799,475,RED,BLACK
  Circle 100,350,300,450,RED,BLACK
  Circle 500,150,650,200,RED,BLACK
  Circle 350,550,450,650,RED,BLACK
return
```

```
InitRobot:
  if Real
    gosub InitDCrobot
  else
    rLocate 400,300
    rInvisible GREEN
    rPen DOWN
  endif
return

Roam:
  dir = 1
  while TRUE
    if random(100)<5 then dir = dir*-1
    gosub CheckForObjects
    if NoObjects
      rForward 40
    else
      rTurn dir*10
    endif
  wend
return

CheckForObjects:
  NoObjects = TRUE
  if Real
    if rRange()<26 // 13 inches
  else
    for angle = -15 to 15
      if rRange(angle)<100 then NoObjects = FALSE
    next
  endif
return

InitDCrobot:
  rCommport PortNum
  rLocate 10,10
  rCommand(MotorSetup,SMALLDC)
  rCommand(SetSpeed,17)
  rCommand(SetReducForwRight,5)
  rCommand(SetMoveTime,34)
  rCommand(SetRotationTime,33)
return
```

Figure 6.11: This program can roam either the real or the simulated robot.

At this point you should be starting to see what programming is all about. Don't worry if some things seem complicated or intimidating. That is true of most everything new. Just study this chapter and experiment with modifying the programs to ensure you understand how they work. Modifying programs in this manner is a lot of fun for many people. If you enjoy it, you might consider a career in engineering or computer science, both of which can involve programming.

When you feel comfortable turn to the next chapter to see how we can make the Ping sensor even more useful.

Chapter 7
Adding a Turret

In Chapters 5 and 6 we learned how to add a Ping ranging sensor to our robot and use it to avoid objects. The sensor was mounted so that it looked straight ahead. Most of the time this is all that is needed, but sometimes it can be nice if the real robot can look to its left and right *without* turning its body (just like the simulator). In this chapter we will add a motorized turret that can turn the sensor itself, greatly simplifying the robot's ability to look to its left and right.

One of the great things about the simulator, as we saw in Chapter 6, is that it has the ability to swivel its simulated ranging sensor on a simulated turret. Let's look at the simple program in Figure 7.1 that demonstrates this capability in an interesting way.

```
main:
  xs=400 // starting position of the robot
  ys=400
  rLocate xs,ys
  SetColor GREEN
  rInvisible GREEN
  rectangle 70,100,300,200,red,gray
  circle 560,50,750,260,red,gray
  for angle = -90 to 90 step 20
    r = rRange(angle)
    x=xs+r*cos(DtoR(angle-90))
    y=ys-17+r*sin(DtoR(angle-90))
    line xs,ys-17,x,y
  next
end
```

Figure 7.1: This program demonstrates the simulator's rotating turret.

There are a few things in Figure 7.1 that need discussion. First, it is important to know that the ranging sensor and its turret are mounted on the front edge of the simulate robot. This is the reason 17 has been subtracted from **ys** in a couple of statements. The variable **ys** represents the center of the located robot and subtracting 17 from that value moves that vertical position in the calculations to just inside the front edge of the robot. The value of angle is used to tell the robot to turn its turret to that position before taking a reading.

The **FOR**-loop creates values for **angle** ranging from -90 to +90. In past programs, the value was incremented by one. In this program the **STEP** option is used to increment the value by 20 so that ten iterations through the loop occur, creating the values -90, -70, -50.....70, 90. A little trigonometry is used to draw the distances measured at the angles used for the measurement. If the math is beyond your current abilities, just accept that it is calculating the end-points of the lines showing the measurements. It is valuable to recognize though, that mathematics is a very valuable tool in engineering, robotics, and programming. The more you learn about math, the better you will be at these professions and the more fun you can have with your robot.

If you run the program of Figure 7.1, it will create the output shown in Figure 7.2. In order to conserve space, from this point on in the book the output from programs will only show the actual images, not the entire output window.

As you can see from the length of the lines in the figure, the robot is measuring distances to objects when they are within view.

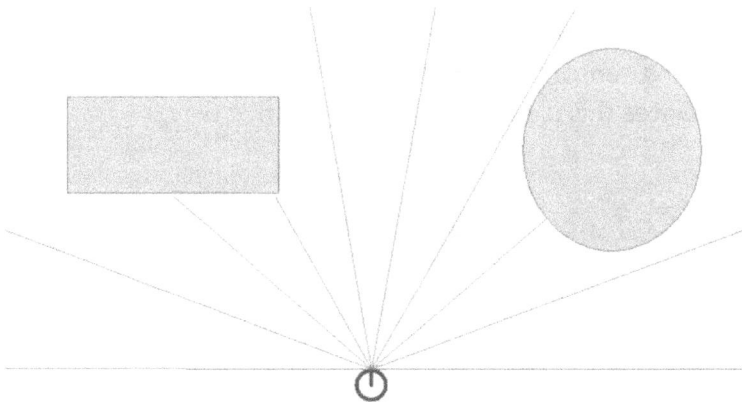

Figure 7.2: The program in Figure 7.1 creates this output.

Adding a Turret to the Real Robot

It would be nice if our real robot had a turret that could rotate the Ping sensor in a similar manner as the simulator. Figure 7.3 shows a Ping ranger mounted on a small micro RC servomotor. A plastic servo-bracket from Parallax makes this easy. The bracket can be glued, screwed, or bolted to the servo horn. A 4-pin header was hot-glued to one end of the servomotor so that it could be physically mounted on the robot by simply plugging it into the breadboard (see Figure 7.4).

Servomotors have three wires. Ground is usually black or brown, the supply voltage is red or orange, and the control line is generally white or yellow. Other than supplying 5 volts and ground to the turret servo, you only need to connect the servo control pin to the RROS pin 10 to complete the setup.

The RROS chip provides all the necessary low-level code to control the turret. You can use the command **rRange(-90)** to look directly left or **rRange(20)** to look 20° to the robot's right just like with the simulator. The turret will automatically move before the reading is taken if it is not already pointed properly. If you have ever programmed a turret mounted ranger, this simplicity should excite you because it gives you more time for application development instead of slaving over low-level code.

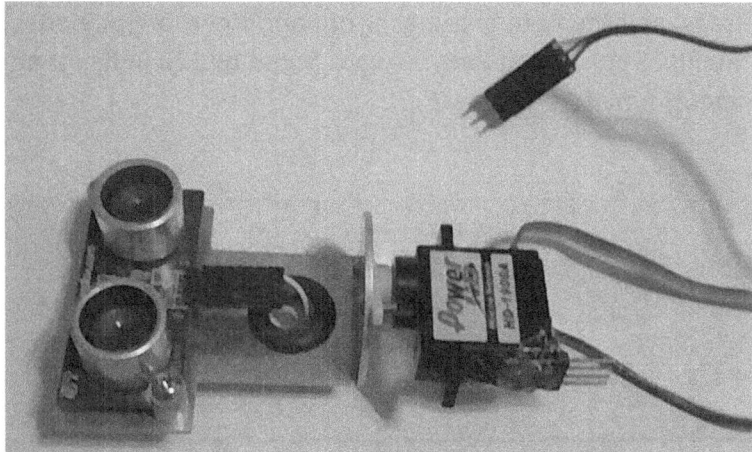

Figure 7.2: A small servomotor can rotate the Ping sensor. Note the 4-pin header hot-glued to the edge of the servo.

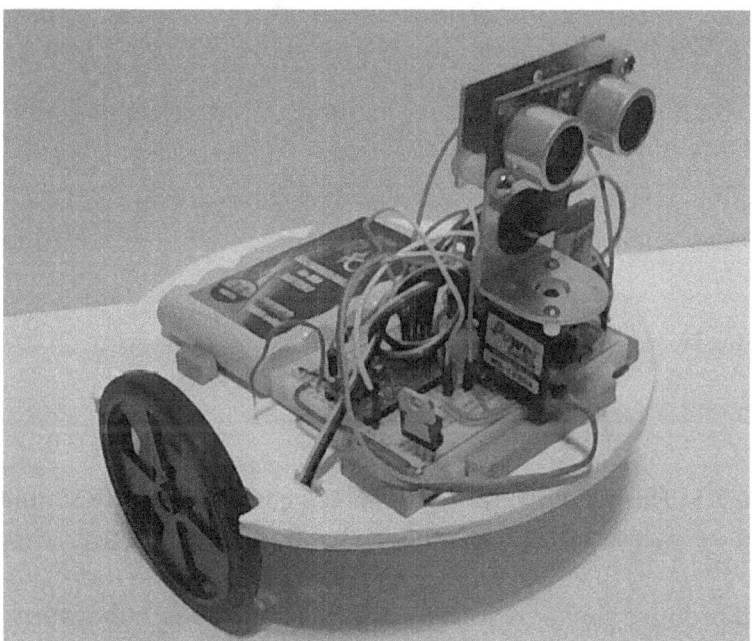

Figure 7.3: The turret is mounted by plugging the 4-pin header into the breadboard (refer to Figure 7.2)

Scanning with the Real Robot

Adding a turret to your robot gives it the ability to execute programs like Figure 7.1. Of course, you will have to initialize your robot properly (either DC or servomotor) and you will not want to draw the objects used in the simulation's environment.

Chapter 7: Adding a Turret

To demonstrate this idea, my real robot was placed on the floor in my office, with a chair and two cases serving as obstacles, as shown in Figure 7.4. When the modified program was run, it produced the output sho in Figure 7.5. Notice the scan shows the chair further away from the robot than the two cases, and the two openings (front and right) are obvious. Note: The maximum range of the RROS controlled Ping is about 5 feet or 120 units. If no object is seen within range, the **rRange()** function returns 255.

Figure 7.4: These real obstacles are placed in the real robot's environment.

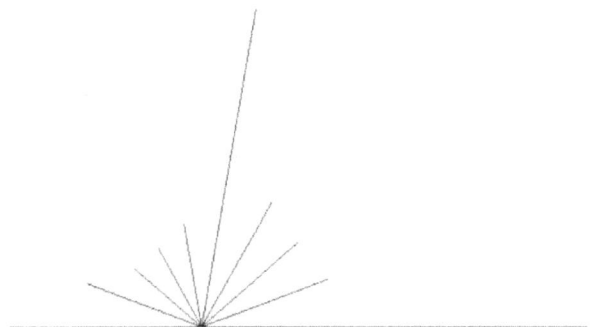

Figure 7.5: This scan diagram (similar to Figure 7.2 for the simulation) shows that it sees objects in its environment.

Of course, you could produce similar results by just having the real robot turn its whole body to take each of the measurement, but using a turret is much easier and faster. Converting the program in Figure 7.1 to work with the real robot is left as an exercise for the reader. At this point in the book, you should feel comfortable with such an assignment. When you feel comfortable with the information in this chapter turn to Chapter 8 and let's see another example of how the turret can allow our robot to perform more exciting behaviors.

Chapter 8
Programming the Robot to Follow a Wall

Now that we have a turret mounted ranging sensor, we can program more exciting behaviors for our robot. In this chapter the robot will be programmed to follow the contours of a wall and in future chapters that ability will be combined with other behaviors to demonstrate how rudimentary machine intelligence can be built in layers.

Using the Simulator for Development
As in previous chapters, RobotBASIC's simulator will be used to develop the algorithm that will govern the robot's behavior. Simulations are widely used throughout industry to allow rapid low-cost experimentation of a wide variety of ideas. Once we have developed the principles of wall-following with the simulation, we will transfer those ideas to a real robot.

Let's start with the program in Figure 8.1. The main section of code performs four tasks by calling subroutines to draw walls to be followed, initialize the robot, find the walls, and follow them. Two of these subroutines are actually completed in the figure.

The **DrawWall** routine uses three circle commands to create a blob-like shape with an exterior edge (wall) that has a variety of interior and exterior curves. The **InitRobot** routine simply locates the simulated robot on the screen and drops a pen so we can see its movements.

The **FindWall** and **Follow** routines have been left blank at this time, but will be developed soon. If we run the program shown in Figure 8.1, it will draw the output shown in Figure 8.2. Examine it to see how **DrawWall** uses circles and ellipses to create the blob mentioned earlier. One of the great things about using a simulation is that it is easy to create a wide variety of environmental conditions for experimentation. Once our wall-following behavior is working, you can easily create a more complex wall for further testing. You will find though, that the variety associated with this shape is complex enough that it serves as a reasonable testing platform.

Notice from Figure 8.2 that the robot has been located a short distance from the wall itself. It is facing the wall though, so it will be easy for us to create the **FindWall** routine. Before you proceed further in this chapter, pause for a short time and think of how *you* would write the **FindWall** routine. Start with deciding *what* you want the robot to do and then think about *how* you can write code to make the robot perform the way you want.

```
Main:
  gosub DrawWall
  gosub InitRobot
  gosub FindWall
  gosub Follow
end

DrawWall:
  Circle 250,200,500,300,RED,RED
  Circle 300,100,550,250,RED,RED
  Circle 350,250,450,350,RED,RED
return

InitRobot:
  rLocate 300,500
  rInvisible GREEN
  rPen DOWN
return

FindWall:
return

Follow:
return
```

Figure 8.1: This code outlines the basic template for our wall-following program.

Figure 8.2: This output is produced by the program in Figure 8.1.

Finding the Wall

So what should the robot do to find the wall? Since we can assume (for this example) that the robot is facing the wall, the behavior needed is to simply move forward until it gets close to the wall. Figure 8.3 shows how readings from the range sensor can be used to determine when to stop the robot.

A **while**-loop continually moves the robot forward in tiny 1-pixel movements as long as the distance to the wall is larger than 12 pixels. Try experimenting with different numbers to control when to stop, but change the program back to 12 when you are finished as a value of 12 seems to stop the robot at a reasonable distance from the wall. Figure 8.4 shows how the robot actually moves to the wall using the code in Figure 8.3. Note that there is no angle give in the **rRange()** statement, so it looks directly forward. This is the same as giving it the parameter **0**.

```
FindWall:
  While rRange()>12
    rForward 1
  wend
return
```

Figure 8.3: This new version of **FindWall** moves the robot forward while it is too far from wall.

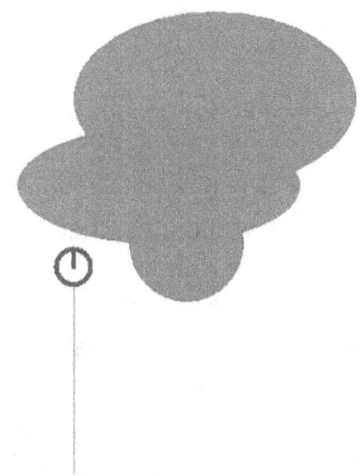

Figure 8.4: The code in Figure 8.3 moves the robot as shown here.

The Real Work

Now that we have a wall drawn, and the robot adjacent to it, we are ready to get down to business and make the robot follow the contours of the wall. Again, start by decided *what* you want the robot to do. A good way to do this is to place yourself in the robot's position. What would *you* do to follow a wall?

Imagine you are blindfolded or in a totally dark room. Assume you have moved forward, until you are close to a wall (just like our robot). How would *you* follow the wall? Think about this before reading further.

Developing the Algorithm

Hopefully you decided that following a wall is pretty easy in a dark room. If someone put you next to a wall in a dark room, you probably would reach out with one hand and use it to stay close to the wall. If you get too close to the wall (your arm is bent), you move away from the wall. If you get too far from the wall (your hand is no longer touching), you move closer. This very simple idea is exactly what we want the robot to do. Figure 8.5 shows a good first approach to solving this problem.

```
Follow:
  rTurn 90
  while TRUE
    if rRange(-90)>50
      rTurn -1
    else
      rTurn 1
    endif
    rForward 1
  wend
return
```

Figure 8.5: This new version of Follow, follows the wall nicely, for a while.

The new version of **Follow** in Figure 8.5 starts by turning the robot 90° to the right in order to generally align it with the wall. An endless **while**-loop checks the distance to the wall by using the range sensor angled 90° to the left. If the robot is further than 50 pixels away, the robot turns back toward the wall. Otherwise, the robot turns away from the wall. After a turn (either toward or away from the wall) the robot moves forward slightly. Note the similarity of this code to you feeling your way along a wall in a dark room as described earlier.

Unfortunately, if you assemble this program and run it, it produces the results shown in Figure 8.6. The robot does indeed begin to following the wall properly. But, as you can see, when the robot encounters an abrupt protrusion in the wall's contour, it collides to cause an error. Part of the problem is that the wall being followed actually turns to the left when the robot needs to be turning right (see the Figure).

As a programmer, the next step is to understand exactly why the algorithm failed and propose one or more possible solutions. The solutions, of course have to be tested to see if they work or have the potential to work (which leads to more proposals and more testing).

Before reading on, examine the code and the output and come up with some ways you think might fix the problem.

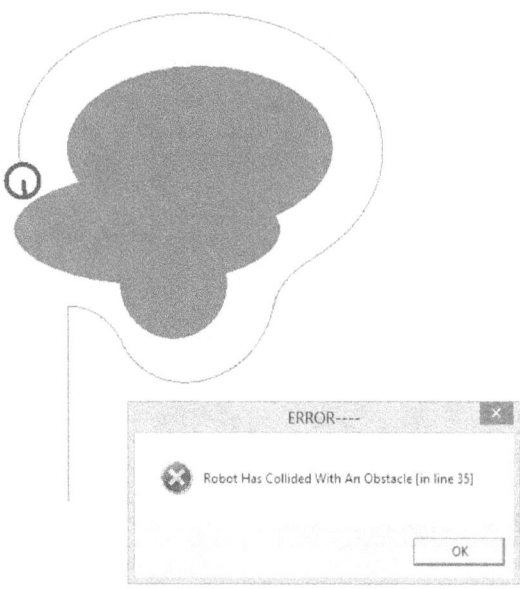

Figure 8.6: The robot follows the wall until it reaches a protrusion.

Possible Solutions

One possible solution is to make the robot follow the wall in the same manner, but to stay further from it. If you increase the distance from 50 to 70 you will see an improvement, but the collision still occurs. If you get to 80 though, the robot is far enough out that it succeeds as shown in Figure 8.7.

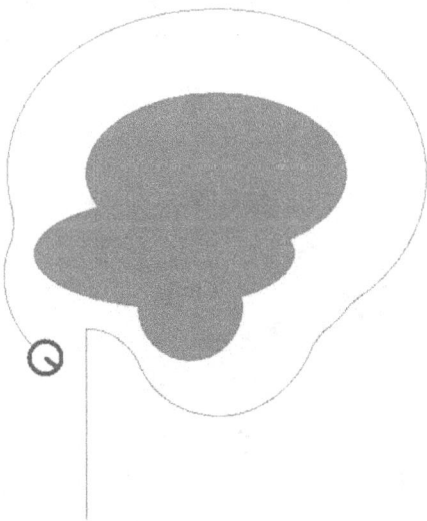

Figure 8.7: Move the robot further from the wall seems to work.

Unfortunately, the fix is only good for this particular wall. If we make the circle obstructing the robot's path a bit wider, as shown in Figure 8.8, the robot fails just like before. Obviously we need a better solution. Can you think of one?

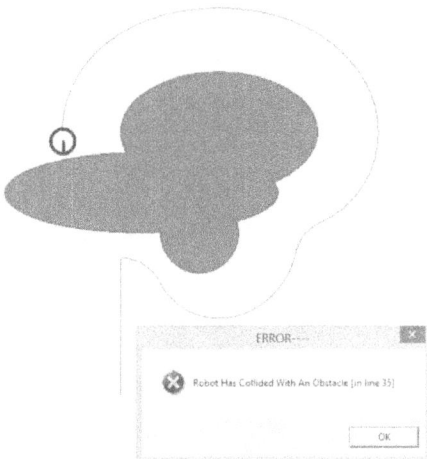

Figure 8.8: Changing the wall results in same error as before.

A Better Solution
What is really needed is for the robot to be looking ahead so it can start turning before it gets too close to the wall. That is *what* that needs to be done. Now we just need to figure out *how* to write code to accomplish it. One way is to have the robot angle its range sensor at about 65 or 75 degrees. It is still looking left enough to keep a steady distance from the wall, but if the wall ahead starts to turn outward, the new angle will give the robot a slightly earlier warning of the upcoming turn. Starting the turn early should help the robot handle problems like the one above.

If you use an angle of $-65°$ for **rRange()** and go back to a distance of 50, you will get the results shown in Figure 8.9. If you compare Figure 8.8 to Figure 8.9 you can see the new algorithm is trying to work. The robot is starting to turn, but just not fast enough.

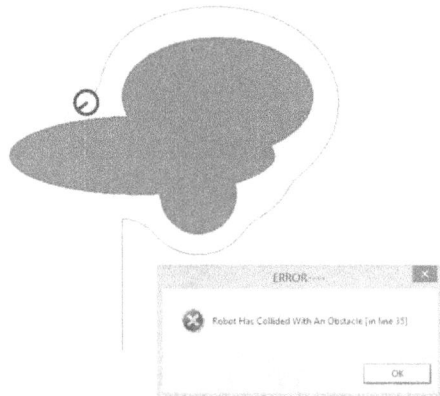

Figure 8.9: If the robot looks ahead slightly it seems to work better.

We can solve the problem shown in Figure 8.9 by telling the robot to make faster turns. The question is, will the robot still perform properly on other areas of the wall. If we change the

rTurn statements to 2 and -2 instead of 1 and -1, you get the result shown in Figure 8.10. This time, the robot almost makes it.

There are several things we can try. We might make the robot turn faster or we could change the angle of the **rRange** statement, or we could increase the distance the robot stays from the wall. You might want to experiment with these options before reading further.

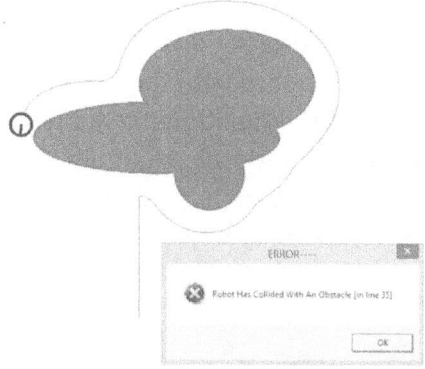

Figure 8.10: The robot is now almost perfect.

A Solution is Found

After experimenting with the parameters mentioned, I found turning ±2 with an angle of 75° and a distance of 60 units seem to work well. Figure 8.11 shows the final output. Experiment on your own and see if you can improve on the robot's wall-following behavior.

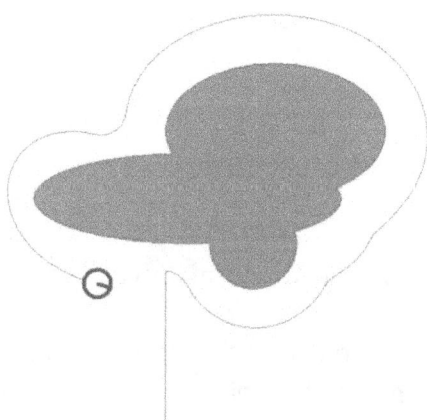

Figure 8.11: The robot seems to handle all situations.

Using the Real Robot

Now that we have the simulation working and have a better understanding of what a robot needs to do to follow a wall, let's port all that knowledge to the real robot. We will be using the DC robot, but as with all our programs, you can use the servomotor robot as long as you initialize it properly.

Chapter 8: Programming the Robot to Follow a Wall

Since the real robot is not exactly like the simulation, we may have to fine-tune some of the parameters that influence the robot's wall-following behavior, but at least we now know what things are important. Various parameters though, are not the only things that have to be considered when porting simulator programs to a real robot.

Real-World RROS Behaviors
One of the differences between the RobotBASIC simulation and a real robot involves their movements. The simulation always moves forward and backward (**rForward**), or it rotates around its center by turning one wheel forward with the other in reverse (**rTurn**). Such motions made it much easier for us to program the simulation itself, but real-world robots do not do well trying to duplicate this action. Let's look at a simple example. Figure 8.12 shows a simple program that moves the simulated robot in a circle as shown in Figure 8.13.

```
rLocate 100,300
rInvisible GREEN
rPen DOWN
while TRUE
   rForward 1
   rTurn 1
wend
end
```
Figure 8.12: This program moves the robot as shown in Figure 8.13.

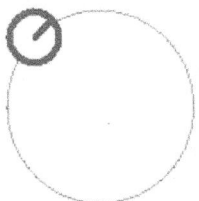

Figure 8.13: The simulation can move in a circular motion.

The important point of Figures 8.12 and 8.13 is that the simulation can move in a circular manner even though it does not have commands for doing so. This curvy motion often happens naturally when the robot is reacting to sensory input. You can easily see this in Figure 8.11 (and others) showing the path taken while executing the wall-following algorithm.

If we tried to execute the program of Figure 8.12 on a real robot though, it would not react as expected. In this particular example, the right wheel would constantly be changing directions. It moves forward when the robot executes the **rForward**, but backward when executing the **rTurn**. Since a real robot has mass and inertia, it will vibrate as the right motor constantly reverses directions.

Chapter 8: Programming the Robot to Follow a Wall

In the real world, a robot should make circular turns by slowing down one wheel or perhaps even stopping it for quick turns (instead of reversing it). It is important to realize though, that sometimes the real robot needs to rotate around its center just like the simulation – but when executing sensory-controlled algorithms, it should not indulge in motions that constantly reverse a wheel's rotation. Perhaps a simple example can clarify this assertion.

If the robot is traveling down a hallway by essentially following one of the walls, it should just slow one wheel when it wants to turn slowly to the right or left to maintain its distance from the wall. If it reaches a doorway it wants to enter though, it then makes sense to *rotate* 90° to face the opening, just like the simulator. Another important point is that when rotating to face the door, the command used will most likely be **rTurn 90**. When moving down the hallway using sensors to constantly adjust its motion, an **rTurn 1** or **rTurn -1** is more appropriate (because the robot should only move a tiny amount before making corrections based on sensor data).

For this reason, the RROS system was designed so the robot always make rotational turns (just like the simulator) if the **rTurn** parameter is larger than 1 or less than -1. If a parameter of 1 or -1 is used, then the programmer has the option of having the robot automatically slow the right wheel when executing **rTurn 1** or the left wheel when executing **rTurn -1**. The programmer can control how much the appropriate wheel should be slowed with the statement **rCommand (SetTurnStyle, 50)**. In this case, the 50 indicates that the slower wheel should be at 50% of the speed of the other wheel. If the parameter is 0, the slow wheel is totally stopped and the robot will rotate around that wheel. If the parameter is 90, the slow wheel is turning almost as fast as the other wheel, causing the robot to have only a slight drift to one side (instead of a distinct turn).

If we are running the program of Figure 8.12 on a real robot with a **TurnStyle** of 50, we can eliminate the **rForward** command from the **while**-loop. Think about it. The simulator needs to rotate, then move forward (over and over) to simulate the circular movement shown in Figure 8.13. We could make it move in a larger circle by using **rForward 2**, or even a huge circle with **rForward 4**. Try making these changes to see how the simulated robot reacts.

If we are using the real robot though, it can move in a similar circle by simply slowing one wheel. It does not need an **rForward** at all, because it still moves forward when one wheel is slowed. If we use a **TurnStyle** of 10, it will move in a fairly small circle. Increasing the size of the **TurnStyle** will automatically increase the size of the circular motion performed by the robot.

Once you understand the above discussion, it is easy to translate simulator behaviors involving circular motions (like wall-following) to a real robot. The program in Figure 8.14 is a translation of the wall-following simulation program to control the real robot. Most changes are identical to that used in previous chapters.

You will notice in Figure 8.14 that some parameters have changed compared to those used for the simulator. Remember, the real robot's distances are in units of ½ inch while the simulation uses pixels. For that reason, the real robot moves until it is 4 units (2 inches) from the wall before it stops when finding the wall. The real robot tries to stay 16 units (8 inches) from the wall while following. These numbers make the real robot react in a similar manner to the simulation.

Notice that the **TurnStyle** has been set to 70 in the **InitDCrobot** module and that the **rForward** has been removed from the **Follow** subroutine. You can control how fast the real robot turns toward and away from the wall by choosing a different **TurnStyle** value. A value of 70 worked well for my robot, but if yours uses different motors or runs at a different speed etc. then you might need to experiment to find a better value.

```
#include "RROScommands.bas"
gosub InitCommands
PortNum = 5 // set to your Bluetooth Port

Main:
  gosub InitDCrobot
  gosub FindWall
  gosub Follow
end

FindWall:
  While rRange()>10
    rForward 1
  wend
return

Follow:
  rTurn 90
  while TRUE
    if rRange(-75)>16
      rTurn -1
    else
      rTurn 1
    endif
  wend
return

InitDCrobot:
  rCommport PortNum
  rLocate 10,10
  rCommand(MotorSetup,SMALLDC)
  rCommand(SensorSetup, PING)
  rCommand(SetSpeed,17)
  rCommand(SetReducForwRight,10)
  rCommand(SetMoveTime,34)
  rCommand(SetRotationTime,27)
  rCommand(SetTurnStyle, 70)
return
```

Figure 8.14: This program makes the DC robot follow a wall.

A Real Wall to Follow

It was easy to create a shape to serve as the wall for the simulation to follow. Now that we are using a real robot, we must have a real wall for it to follow. You could use a makeshift wall made from furniture, boxes, briefcases, etc. but Figure 8.15 shows a better solution.

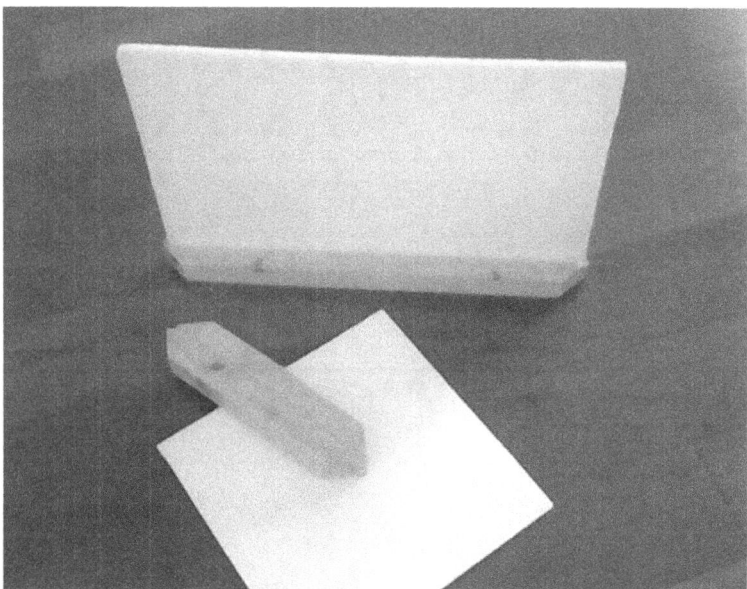

Figure 8.15: Wall modules make building real walls and obstacles easy.

The wall sections in Figure 8.15 are composed of two parts. The base of each section is a short piece of 1x2 lumber with a slot cut in it to accommodate a piece of foam board. This wood base provides the weight to give the wall section stability. If you don't have a way to cut the slots, you could just glue the foam board to the wood base. If you choose gluing, I suggest you use the thicker foam board (about ½ inch compared to the typical ¼ inch) to give you a larger area for the glue. The pointed ends of the wood bases allows more flexibility when placing the sections to form a wall (or obstacles) as shown in Figure 8.16.

Figure 8:16: Curvy walls are easy to build with the sections shown in Figure 8.15.

Build a wall as you see fit and test the program in Figure 8.14 to verify it can make your real robot follow it. Adjust pertinent parameters appropriately if the robot fails to follow the wall. Try experimenting with all the options to ensure your robot can follow walls with a variety of curves.

Chapter 9
Adding a Line Sensor

The wall-following discussion in Chapter 8 is fairly complex, but it provides a lot of information. For that reason, it is suggested you study it carefully before moving on. When you are ready, let's add a new type of sensor to our robot.

Figure 9.1 shows a QTR-1A reflective sensor from www.Pololu.com. It emits a low-level infrared light and can detect if that light is reflected off a nearby object. For our use, it differs from a ranging sensor in two ways. First of all, it can only determine if something is reflecting the light back or not. It cannot determine how far away the object is. Second, it is designed for very short range situations. We will mount this sensor under our robot about 1/8 inch from the floor. If our robot is over a white surface such as a poster board, the RROS will communicate with the sensor and not report that it sees anything. If it is over a dark area such as black electrical tape then the RROS will reflect that.

Figure 9.1: The QTR-1A sensor from Pololu can detect lines.

We can mount the sensor on our real robot as shown in Figure 9.2. The QTR-1A comes with two sets of pins (one straight and one at a right-angle) that can be soldered to the sensor itself. This allows you to use a short cable (male pins on one end and female on the other) to connect the sensor to the robot's breadboard. Alternatively, you could just solder three wires to the interface holes on the sensor. How to connect these wires will be discussed shortly.

Mounting the Sensor
The sensor itself can be hot glued to a piece of foam board that has been glued to the bottom of the robot (see Figure 9.2). You want the sensor to sit about 1/8 in above the floor. The sensor will not work properly if it is too close or too far away. Make sure you test the distance to the

floor before gluing. In my case, the sensor was slightly too far from the floor so I added a thin piece of wood to adjust the distance (as shown in Figure 9.2). The wires from the sensor should run through a hole cut in the base to the breadboard. Make sure the wires you use are long enough to reach their connecting points.

Electrical Connections

Connections to the breadboard are easy. The ground pin (GND) should connect to the breadboard ground and the center pin (power) on the sensor should connect to the 5V supply that was used for the Bluetooth transceiver. Finally, the OUT pin should be connected to pin 5 on the RROS chip. That is all there is to it. Your sensor should be ready to use.

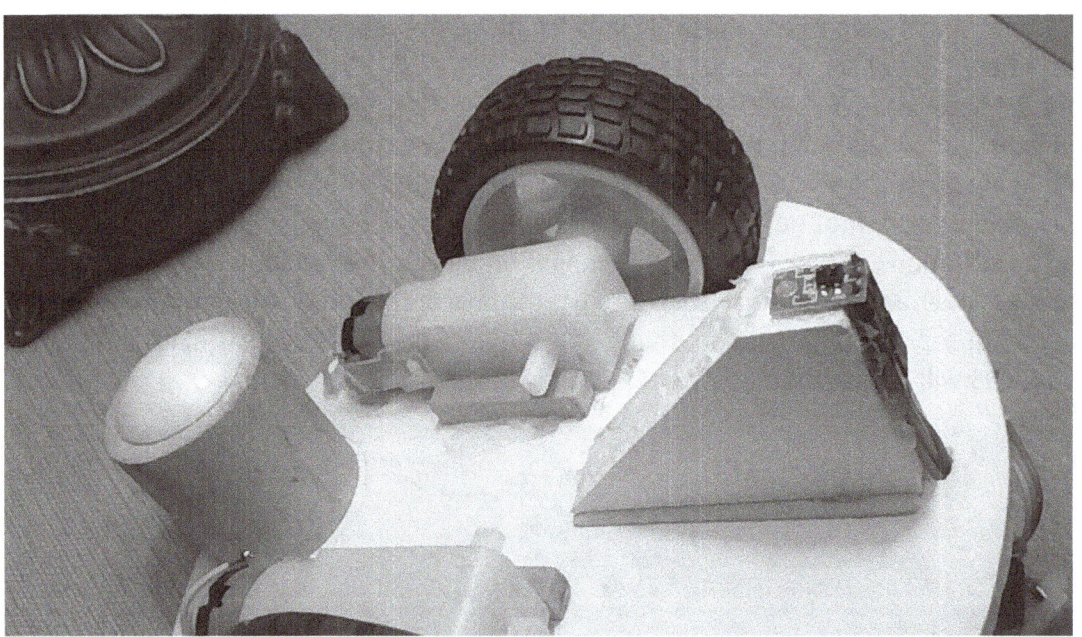

Figure 9.2: The QTR-1A sensor can be hot glued to a foam board wedge.

The RROS interfaces with the sensor and reports the data received to RobotBASIC using Bluetooth. Your programs can determine if the sensor sees a line using the function **rSense()**. The value of this function will be zero if no line is seen and 1 when the sensor is over a dark (non-reflective) area. The program in Figure 9.3 can be used to test to see if your sensor is working.

An Example Program

The main program in Figure 9.3 continually uses an **IF-ELSE-ENDIF** to check for a line and prints a message on the screen. In this case, **xyString** commands are used instead of **Print** statements. The **print** command always prints on the next line down on the screen and is very simple to use, especially for beginners. The **xyString** requires that you tell it where to print, in this case at pixel position 100,100. Since the lines to be printed are exactly the same size, printing one of them in the same place will erase the other.

The **rForward 0** command in Figure 9.3 deserves mentioning. The ranging data used in previous chapters is updated when the **rRange()** function is executed. Some sensors, including line sensors are only updated when the robot is told to move with either **rForward** or **rTurn**. IF you think about it, this makes perfect sense. If the robot is sitting on or near a line, the reading from the line sensor should not change unless the robot moves. Having the line sensor data update automatically when movements are requested, is not just convenient, it is actually faster as well.

```
#include "RROScommands.bas"
gosub InitCommands
PortNum = 5 // set to your Bluetooth Port

Main:
  while TRUE
    rForward 0
    if rSense()=1
      xyString 100,100,"A line is seen "
    else
      xyString 100,100,"No line is seen"
    endif
  wend
end

InitDCrobot:
  rCommport PortNum
  rLocate 10,10
  rCommand(MotorSetup,SMALLDC)
  rCommand(SensorSetup, PING)
  rCommand(SetSpeed,17)
  rCommand(SetReducForwRight,10)
  rCommand(SetMoveTime,34)
  rCommand(SetRotationTime,27)
  rCommand(SetTurnStyle, 70)
return
```

Figure 9.3: This program can test your line sensor to see if it is working properly.

Testing the Line Sensor
Prepare a piece of white poster board with a black line on it (preferably using black electrical tape for the line). Sit your robot on the poster board and run the program. Manually move the robot over the line and off again while watching the display. It should tell you when the robot is over the line. If it does not work properly make sure your wiring is correct and that you are using a non-reflective *dark* line on a white surface and that the sensor is 1/8 inch from the floor.

Developing a Line Following Behavior

At this point we are ready to develop a program to make the robot follow a line. We will start by using the simulator to develop the algorithm. The simulated robot actually has three line sensors near its front edge, compared to the one sensor we mounted on our real robot. Three sensors allows many more options that make it possible to write programs that can follow a line in a more sophisticated manner. Such programs, however, are not suitable for this beginner's book due to their complications. For that reason we will only use a single line sensor in our programs (even the RROS supports three line sensors). This will certainly put limitations on our robot's ability to follow lines, but overall, we should be able to create a reasonable line-following behavior. Readers that wish to further their studies can add additional line sensors to their robot in the future (refer to the RROS Users Manual).

The Algorithm

The line-following algorithm for a single line sensor will be very similar to the wall-following algorithm used in Chapter 8. Rather than trying to keep the robot centered on the line (as is possible with multiple line sensors) we will make our robot hug the edge of the line in much the same way as the robot hugged the wall it followed.

Basically, if the robot is on the line, we will have it turn right, which should take it off the line. If the robot is not on the line, it should turn to the left, which should make it move back on the line. Notice the similarity to the wall-following behavior from previous programs.

Figure 9.4 shows a program that will demonstrate this algorithm with the simulated robot. It starts by calling a subroutine to draw the line to be followed. The line drawn is green, so the **InitRobot** module needs to specify that green is an invisible color so the robot can travel over the line without seeing it as an obstacle. Notice also that the line is 20 pixels wide. The speed of the robot is slowed down after the robot is located so it is easier to see what it is doing as it follows the line.

The actual line-following is implemented inside of **Main**, inside a **while**-loop. An **if**-statement checks to see if the right-hand line sensor (of the 3 sensors on the simulator) is on or off. It does this by using the **&1** which actually performs a binary **AND** function. A full discussion of binary math is outside the scope of this text, so just accept that using **rSense()&1** is how you test just one of the simulator's three line sensors (the one connected to RROS pin 5 in this example).

If the line sensor is **1** or **TRUE**, meaning a line is seen, the robot turns to the right with **rTurn 1**, otherwise, the robot turns left with the **rTurn -1**. After making the turn (either left or right) the robot moves forward with the **rForward 1**. Hopefully you see how this implements the algorithm described earlier.

```
Main:
  gosub DrawLine
  gosub InitRobot:
  while TRUE
    if rSense()&1
      rTurn 1
    else
      rTurn -1
    endif
    rForward 1
  wend
end

InitRobot:
  rLocate 50,500
  rSpeed 2 // slow robot down
  rInvisible GREEN
return

DrawLine:
  LineWidth 20
  SetColor GREEN
  line 50,500,50,450
  lineto 75,400
  lineto 100,375
  lineto 150,350
  lineto 200,340
  lineto 250,350
  lineto 300,350
  lineto 350,330
  lineto 400,320
  lineto 450,300
  lineto 500,270
  lineto 550,200
  lineto 600,180
  lineto 600,100
return
```

Figure 9.4: This program allows the simulated robot to follow a line.

If we run the program, we will get the output shown in Figure 9.5. Notice how the robot hugs the right side of the line as it moves. Can you explain why? Can you modify the program so the robot hugs the *left* side of the line?

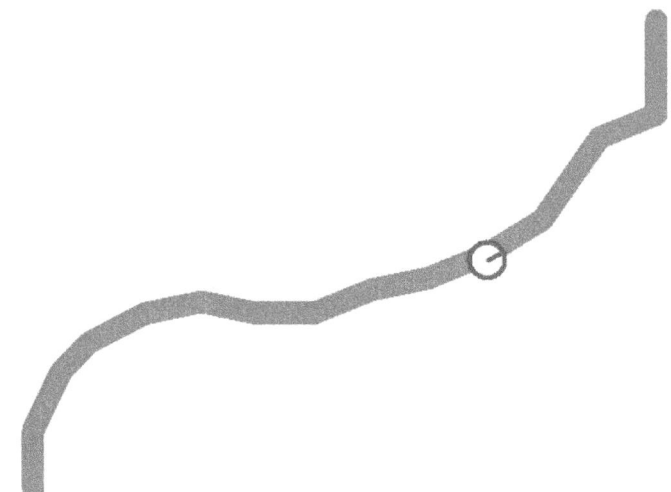

Figure 9.5: The robot follows the line by hugging its right side.

The problem is that the robot fails when it gets to the sharp left turn in the line near the end. Figure 9.6 shows the details of what happens. Can you explain why it happens?

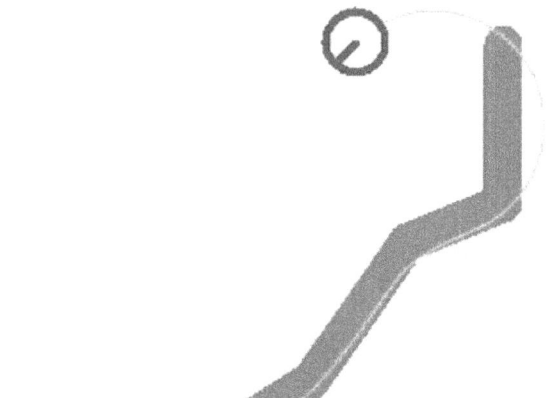

Figure 9.6: The robot loses the line when it turns sharply.

When the line turns sharply to the left, the robot tries to turn left to find the line again. Unfortunately, the robot does not turn fast enough so it ends up too far to the right of the line. This makes it approach the line at an angle that is too sharp for it to acquire the line again. Look at Figure 9.7 which show the same situation, but with the robot continuing on with it attempt to find the line. Notice that when it crossed over the line, the robot turned right just as the algorithm dictates. This is especially visible at the last point where the robot crossed the line. Turning right is obvious, but also obvious is that it does not have enough time to get parallel with the line. If the line were wider, much wider, the robot would have more time to turn.

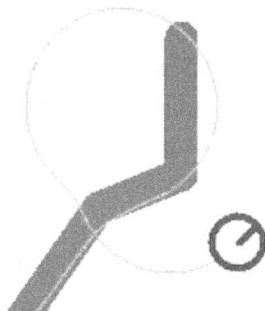

Figure 9.7: The robot struggles to find the line again.

If we widen the line to 50 pixels, the robot does indeed acquire the line again as shown in Figure 9.8. In this example, it is very easy to see how the robot is trying to turn right whenever it is on the line. This is one example of the limitations of having only a single line sensor. With more sensors the robot could use the additional data to start turning faster when the line is almost lost. This is difficult (but not impossible) to do with a single sensor. With more programming experience, both situations can be solved, but for now we will just examine the program shown.

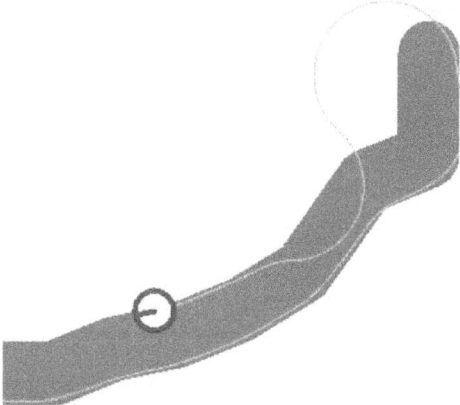

Figure 9.8: Changing the line's width alters the robot's path.

We have seen what happens with wider lines. Figure 9.9 shows the robot's reaction to a narrow line. Notice that with a thinner line, the robot loses the line on a less sharp turn.

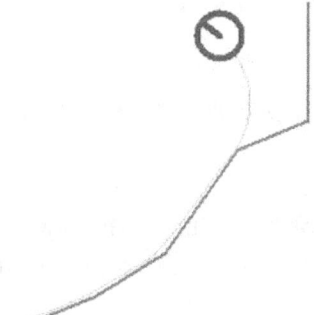

Figure 9.9: The robot loses the line earlier if the line is thin.

Porting to the Real Robot

The point of this discussion is that it helps you understand exactly how the robot responds to different situation. And this knowledge will be valuable when we start programming the real robot to follow a line. Figure 9.10 is a translation of the program in Figure 9.4 for the real robot. Perhaps the first thing you notice is how simple this program is because it does not have to draw the environment. The code for the actual algorithm is nearly the same. The only difference is that the **rForward 1** has been omitted. The reason of course, is that the **TurnStyle** has been set to 70 just like in the wall-follow programs of Chapter 8, and with this **TurnStyle** the robot moves forward as it turns. Of course, your robot could be significantly different from mine, so adjust your robot's speed and **TurnStyle** appropriately if you have problems.

```
#include   "RROScommands.bas"
gosub InitCommands
PortNum = 5 // set to your Bluetooth Port

Main:
  gosub InitDCrobot:
  while TRUE
    if rSense()&1
      rTurn 1
    else
      rTurn -1
    endif
  wend
end

InitDCrobot:
  rCommport PortNum
  rLocate 10,10
  rCommand(MotorSetup,SMALLDC)
  rCommand(SensorSetup, PING)
  rCommand(SetSpeed,17)
  rCommand(SetReducForwRight,10)
  rCommand(SetMoveTime,34)
  rCommand(SetRotationTime,27)
  rCommand(SetTurnStyle, 70)
return
```

Figure 9.10: This program makes the real robot follow a line.

A Real Robot Following Real Lines

Figure 9.11 shows the real robot on a real line. The line was reasonably wide, so the robot performed as expected. Figure 9.12 shows the robot trying to follow a thin line. As expected, it lost the line on even moderately sharp turns. If you've understood this chapter, you should know how you might improve on the robot's ability to follow thin lines. Test your ideas with your robot and various lines sizes. Try adjusting the value for **TurnStyle** to improve performance. Increase the parameter if you want the robot to slower and decrease it to turn faster.

Figure 9.11: The robot easily follows a wide line.

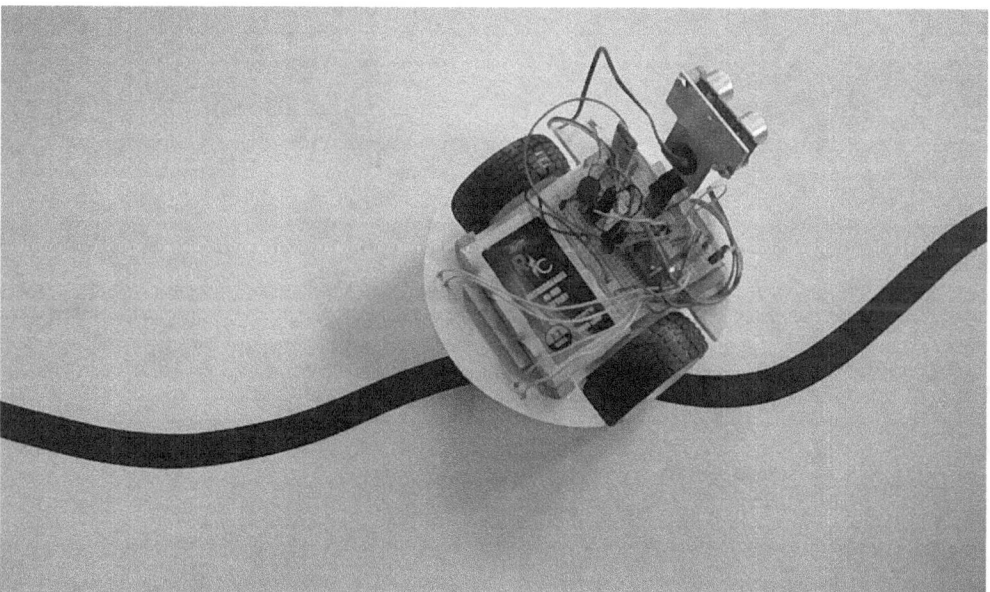

Figure 9.12: The robot loses the line quickly if it is too thin.

In Chapter 12 we will combine line-following and wall-following behaviors to make a robot handle a reasonably complex problem.

Chapter 10
Avoiding Objects on a Line

The real secret to programming is that small modules can be combined to build bigger ones and then the bigger ones can be combined to build even bigger ones etc. When this continues, the end result becomes huge programs with massive capabilities. This book, even though it is for beginners, would be remiss if it did not demonstrate how combining simple behaviors can actual build a complex program capable of performing a complex task.

A Complex Task
In this chapter, we want the robot to follow a line until some obstacle blocks the robot's way. When this happens, the robot should abandon its line-following behavior and use wall-following to find its way around the obstacle. But, it should only follows the contour of the object until it sees the line again, at which time it should return to the line-following behavior. Figure 10.1 shows what we want to happen.

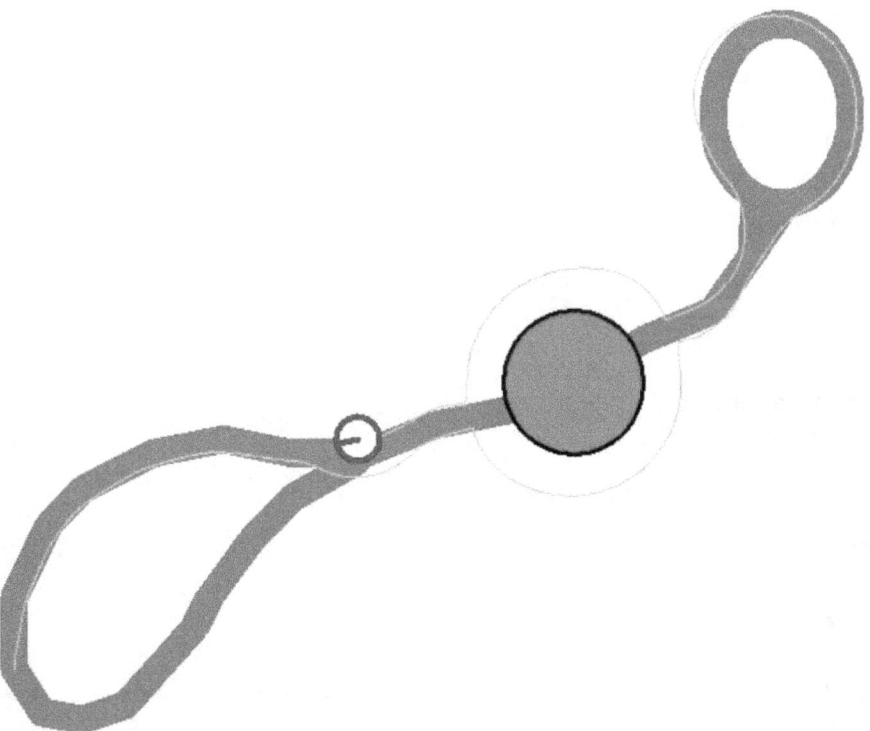

Figure 10.1: The robot avoids objects blocking its path on the line.

The robot starts in the lower left hand corner and moves along the line until it comes to the object blocking its path. It goes around the object and continues following the line. But notice, the ends of the line are now circular so that the robot can follow around and continue following the same line, but on the upper side (still the robot's right side).

When the robot gets back to its starting position it takes a different path back to the obstacle and again avoids it. From then on, the robot stays on the outside of the line and switches back and forth between line-following and wall-following behaviors, as needed. A cool aspect of this program is that the robot never stops because the lines curve back on themselves. The program that produces this output (but without the line showing the robot's path) is provided in Figure 10.2.

```
Main:
  gosub DrawLine
  gosub InitRobot:
  gosub FollowLine:
end

FollowLine:
  // a modified line follow
  while TRUE
    if rRange()<10 then gosub AvoidObject
    if rSense()&1
      rTurn 1
    else
      rTurn -1
    endif
    rForward 1
  wend
return

AvoidObject:
  // basically a modified wall follow
  rTurn 90
  while not rSense()&1
    if rRange(-75)>35
      rTurn -1
    else
      rTurn 1
    endif
    rForward 1
  wend
  rForward 20
  rTurn 90
return
```

```
InitRobot:
  rLocate 50,500
  rSpeed 2 // slow robot down
  rInvisible GREEN
return

DrawLine:
  LineWidth 20
  SetColor GREEN
  line 50,500,50,450
  lineto 75,400
  lineto 100,375
  lineto 150,350
  lineto 200,340
  lineto 250,350
  lineto 300,350
  lineto 350,330
  lineto 400,320
  lineto 450,300
  lineto 500,270
  lineto 550,250
  lineto 600,180
  lineto 600,100
  circle 550,50,650,175
  line 580,170,578,210
  line 50,500,70,530
  lineto 100,535
  lineto 130,525
  lineto 180,470
  lineto 190,450
  lineto 220,410
  lineto 250,380
  lineto 295,355
  LineWidth 3
  circle 400,250,500,350,BLACK,RED
return
```

Figure 10.2: This program allows the simulated robot avoid objects while following a line.

The Differences

Most of the program in 10.2 is the same or very similar to what you have seen in previous chapters. Other than the **DrawLine** module, there are really only slight differences in two modules. Let's look first at the module **FollowLine**. The only change (when compared to the **FollowLine** from previous chapters) is shown in **bold** in the figure.

The new **if**-statement monitors the ranging sensor and calls **AvoidObject** when an object is detected. **AvoidObject** is basically a modified wall-follow subroutine. It has several changes when compared to the original. All three changes are in **bold** to help you locate them.

The first bold line is the **while**-statement. In the previous version, this was a never ending loop with a controlling parameter that was always **TRUE**. In this version, the loop only continues as long as the robot does not see a line. This simply means the wall-following behavior will continue until a line is found (meaning the robot has found its way around the object). When the wall-following behavior is terminated, the robot needs to align itself with the line so that the line-following behavior can resume. It does this by moving forward slightly (to get on the line) and then turning 90^0 to the right to complete the alignment.

Easier Than Expected
Hopefully you are starting to feel pretty good about your programming skills. At this point you are probably able to read through the program in Figure 10.2 and actually understand how it works. You may not be comfortable yet about developing your own programs, but the first step is being able to read and understand what others have written.

Actually, your next step should be to experiment with the programs in this book. Study the code and predict what would happen if you make various changes, then make the changes and test the program to see if you were right. You can also make changes to see if you can make a program better. Try simple things like modifying the shape of a wall or line that the robot will follow, then move on to experimenting with the controlling parameters to see what makes the program worse or better.

A Challenge
At this point in the book, you should understand all the algorithms that have been discussed, and feel fairly confident at converting a program developed for the simulator so that it can control the real robot – after all, you did this with a simple program back in Chapter. To help you build your skills I want to leave you with a challenge. The program in Figure 10.2 is composed almost entirely of code discussed in detail in previous chapters. With the additional explanations in this chapter, you should have a good understanding of how each of the new routines work. Additionally, you now have experience converting simulator programs so that they work with a real robot.

Your challenge is to convert the program of Figure 10.2 so that it can control a real robot. Study the previous examples if you have problems and experiment with various parameters (also as discussed in previous chapters) to fine-tune your robot's performance so that it works to your satisfaction. When you are successful, turn to Chapter 11 to see a variety of options for improving your programming skills and knowledge in the future.

Chapter 11
What's Next?

Free and Mostly Free
One easy and totally free option for continuing your studies is to read the RobotBASIC help file. It provides hundreds of pages describing the commands and functions available through RobotBASIC. You will quickly find that the material in this book barely scratches the surface of what RobotBASIC can do. On your first read through the file, don't try to understand every detail. Instead, try to discover the big picture by learning the types things RobotBASIC is capable of. Later when you are writing your own programs, you only need to remember that there was a command or function that can help you accomplish your goal. At that point you can look up the details and write some test programs to verify your understanding of the specifics before actually trying to use something new in a big program.

Another great way to study coding is to read through some of the example programs provided in the free download of RobotBASIC. You might also want to watch some of the many YouTube videos that provide information on RobotBASIC's capabilities.

If you are really interested in Robotics, consider subscribing to Servo Magazine. It is the premier publication for hobby robotics. I personally have published dozens of articles demonstrating various aspects of RobotBASIC and subscribers have instant access to historical issues so you will be able to read them all.

The RROS system can support more sensors and capabilities that were covered in this text. Your robot can have perimeter proximity sensors, battery monitoring, an electronic compass, a beacon detector, and more. The best way to learn about these upgrades is to download a free PDF of the *RROS User's Guide* from the RROS tab at www.RobotBASIC.org.

Books on RobotBASIC
Hopefully you are able to find everything you need or want to know from the abundance of free resources for RobotBASIC. If you want to delve even deeper into some particular subject there are an abundance of books available to provide the information. If you can achieve your goals with our free materials, that's great. If you decide to buy some of our books to help support our efforts to provide support for RobotBASIC, then we thank you. Below are general descriptions of some of the books available on Amazon.com. Most are available in print and as low-cost ebooks. Schools can contact me directly through www.RobotBASIC.org for quantity discounts.

If you look up any of these books on Amazon, you can view a detailed description of their content, access reviews from RobotBASIC users, and read sample text to help you decide if the book might be what you want or need.

These are only some of our publications and the list is constantly growing. Search Amazon.com to see everything that is currently available.

 The RROS Manual is free for download on the RROS tab at www.RobotBASIC.org, but a printed version is preferred by many readers. It provides details of everything the RROS can do as well as many example programs. It is not written for beginners, but supplies a massive amount of information about building RobotBASIC robots using the RROS.

 RobotBASIC's help file is integrated into the language, but a printed version is available to those wanted the convenience to read it anywhere at any time. You could print out the help file, but it will need a lot of reformatting and probably use an entire ink cartridge.

 If you want to learn more about hardware and how sensors and motors can be controlled from RobotBASIC programs, then this is a good place to start. You will also be introduced to vision, speech, voice recognition and interfacing with microprocessors.

 This book gives you a glimpse into how the RROS system works internally. Many of the techniques used to enhance the Pololu 3pi robot made their way into the RROS system.

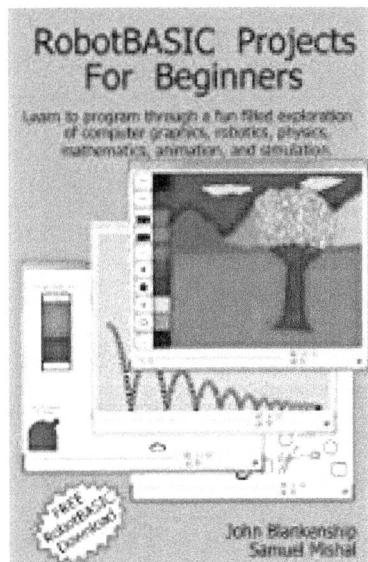 This book will teach you about programming by writing simple animations, simulations, video games, and other graphical applications. If you are a beginner and want to learn about programming, this book is our number one recommendation. The projects will make you love programming.

 This book teaches you about programming like the above book, but all of the example programs are robotic related.

 This book provides more details about developing algorithms for robotic behaviors than any RobotBASIC publication. Nearly everything is done with the simulator so learning how to program a robot will never be easier.

 This is one of our most advanced books and definitely not for beginners, but if you really want to learn how to build a real robot, this is for you. Arlo is a mobile robot with two arms, vision, speech, voice recognition, an animated face, and much, much more.

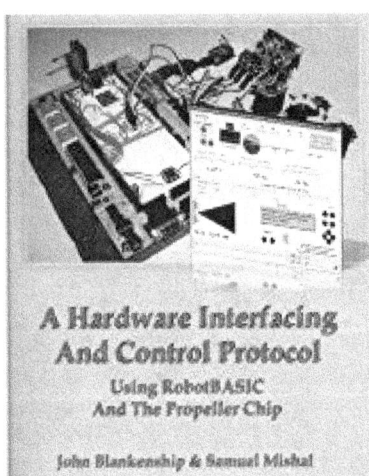 This another of our advanced books. It provides programs that interface with a variety of sensors and Parallax's Propeller processor. If you are a Propeller fan, you need this book.

Hopefully you realize there are many ways to proceed. Learning to program can be fun and exciting but it also develops critical thinking and problem solving skills that can carry over into many aspects of your life. It can help prepare you for careers in programming, engineering, and mathematics. If you love science and technology, you should learn to program.

Appendix A
Breadboarding

Since readers of this book are probably novice robot builders it is important to have a way to build electronic circuits easily without soldering. Using a solderless breadboard as shown in Figure A.1 is a great solution.

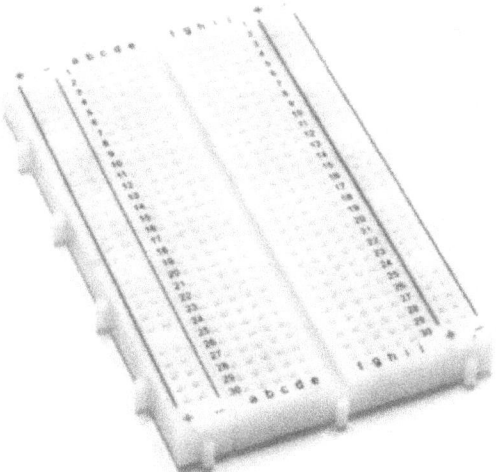

Figure A.1: Solderless breadboards make constructing circuits easy, even for beginners.

Solderless breadboards come in a variety of sizes, both larger and smaller than the one shown in Figure A.1, which is a 400 point board with measurements slightly larger than 2x3 inches. This size board is perfect for the chassis discussed in Chapter 2. You can find boards like this by searching eBay for **solderless breadboard**. Before we mount the board on our robot though, let's learn how to use the board to construct a circuit.

On this particular board, there are 60 sets of 5 holes arranged in 2 columns. Each column has the sets of 5 holes labeled from 1 to 30. The 5 holes in the left column are labeled a,b,c,d, and e. On the right column they are labeled f,g,h,i, and j.

You can connect two wires (or other components) together by plugging them into appropriate holes on the board. Look at Figure A.2. Notice the 5 holes in left column, position 1, are circled. All of these holes are connected together inside the board which means if we plug 2 wires into any of these 5 holes, the 2 wires will be connected together electrically. The same is true of any set of 5 holes on the board.

Notice that there are 10 sets of 5 holes at the top of the board shown in Figure 3.2. There is a red line and two + (positive) signs associated with the top row and a blue line and two –

(negative) signs for the bottom row. This pattern is duplicated at the bottom of the board. Usually, all five sets in these rows are connected together (as a 15 hole set) and are often called a *bus* line. Typically these rows of holes are used for the positive and negative battery voltage for your robot. As an example, the top two rows might be used for the main battery voltage, which might be 7.2 volts and the bottom two rows might be used for a 5 volt supply (which is how the robots in this book are wired). We will talk more about this shortly, but the idea is that when some part of your circuit needs a specific supply voltage, you can connect it to any of the holes making up the bus used for that voltage.

It is worth mentioning that some boards might have the bus lines broken in the middle giving you the possibility of having 8 bus lines instead of the 4 discussed here. Again, we will address this subject shortly.

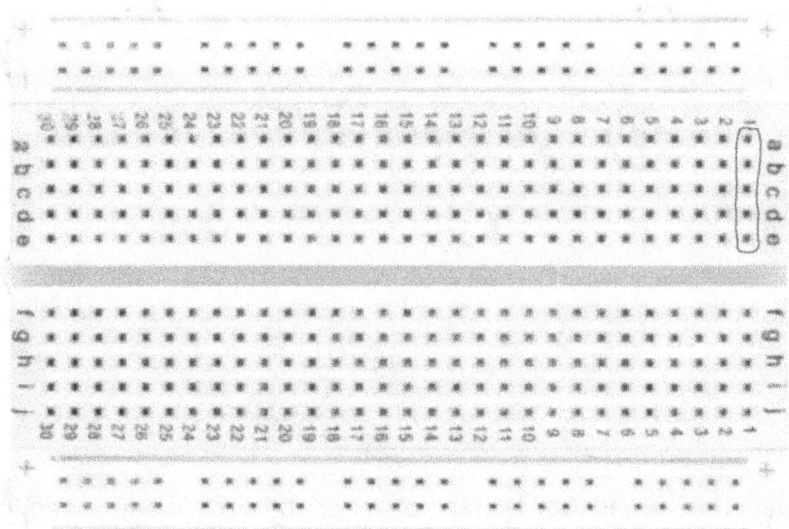

Figure A.2: Each set of five holes on a breadboard are connected together.

In order to help you understand how to build circuits with a breadboard, let's look at a specific example. Look at the circuit schematic shown in Figure A.3. This circuit is not anything real so don't try to actually build it. Its only purpose is to demonstrate how to build circuits from schematics. The schematic shows a RROS chip which has 24 pins ordered as shown. Pin 24 of the RROS chip is connected to a positive battery terminal. The negative battery connects to a symbol called *ground*. Typically there will be many items in a schematic that connect to ground, and additional ground symbols will be used instead of drawing a lot of overlapping wires (which could be confusing).

The positive battery is also connected to a resistor (see the symbol labeled R). The other end of the resistor connects to ground and to the + terminal of a capacitor (labeled C). Some capacitors are not polarized and the ends can be interchanged just like resistors. Electrolytic capacitors have their leads labeled + and - and must be connected as shown in the schematic.

Finally, the negative lead of the capacitor (most schematics only label the + lead) connects to pin 23 on the RROS chip. Throughout this book, schematics will be used to show you what needs to be built. You will need to understand the schematic and build its equivalent circuit on a breadboard.

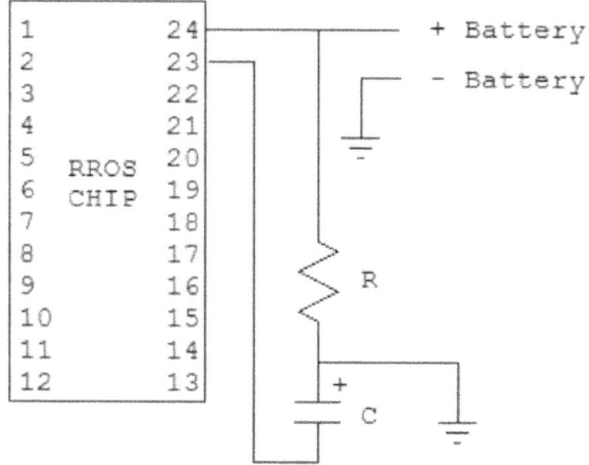

Figure A.3: The only use for this schematic is to teach you how to use a breadboard.

Figure A.4 shows one way to build the schematic of Figure A.3 on a breadboard. As you will see, there are nearly an endless number of ways to actually build this circuit – there is NOT one *right* way.

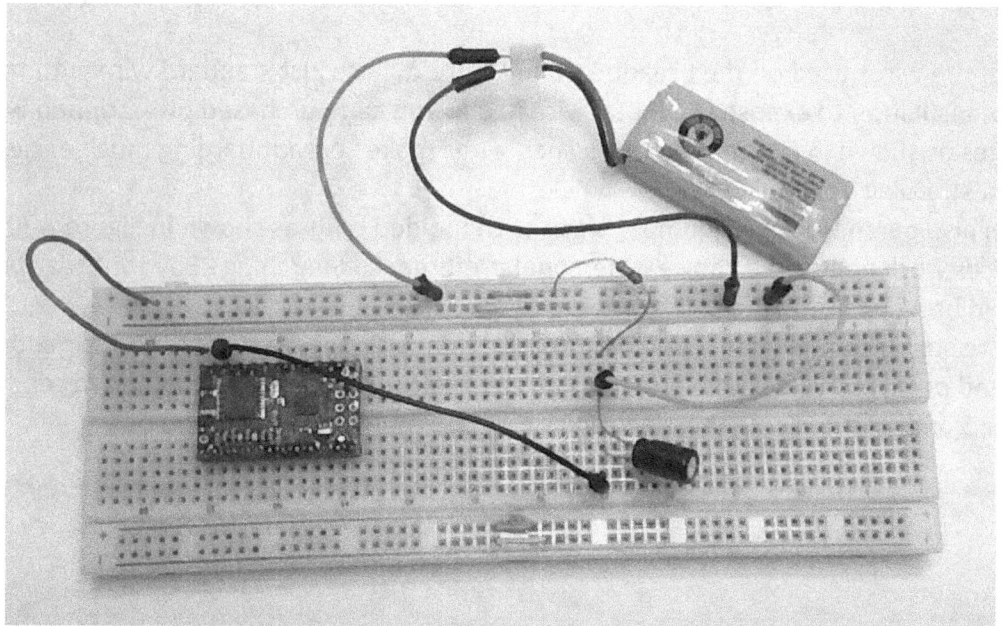

Figure A.4: This is one way build the schematic shown in Figure A.3.

Perhaps the first thing you noticed about Figure A.4 is that the breadboard is much larger than the one shown earlier. All solderless breadboards are similar so this one will work fine for demonstrating how to build a circuit. Let's look at the details of this construction.

The battery wires are connected to the positive and negative bus lines at the top of the board. Since this board is larger, notice that it has 10 sets of 5 holes making up each bus. Actually, as mentioned earlier, each bus on this board is actually 2 buses, each made up of 5 sets of 5 holes. The 5 sets on the left are connected together (as are the 5 sets on the right), but the left side is NOT connected to the right side. Notice that 4 wires (one on each bus line) connect the left and right sides of each bus making the buses on this board work exactly like the smaller board shown earlier.

Notice there is a wire from pin 24 on the RROS chip to the positive bus line, connecting it to the positive battery terminal, just like in the schematic. Notice also that the negative bus line can be thought of as ground, so any ground point in the schematic can be connected to holes making up this bus.

In the schematic, one end of the resistor was connected to the positive battery. In Figure A.4, a resistor is plugged into a hole in the positive bus line. The other end of the resistor plugs into one of the 5-hole groups on the main board area. If you look at the schematic of Figure A.3, you will see that this end of the resistor connects to ground and to the positive end of a capacitor (if you could read the capacitor you would see that the leads are labeled + and - . Look at Figure A.4 and verify that the positive end of the capacitor also plugs into the same 5-hole group used by the resistor. Notice also that a wire connects this same point to the ground bus.

Finally, the negative end of the capacitor is connected to pin 23 on the RROS chip. Study the circuit and the schematic carefully until you see clearly how the breadboard is used. It may seem complicated, but once you get used to it, you will find it a convenient way to quickly construct circuits.

Some of the wires used on the breadboard in Figure A.4 are just standard wire with the ends stripped of insulation to expose the metal wire itself. You can purchased pre-stripped wires of various sizes as shown in Figure A.5. Using such wires makes breadboarding much easier and faster than stripping your own wire.

You can also purchase special jumper wires with molded ends as shown in Figure A.6. Both type of wires work fine so the choice is personal preference. Searching eBay for **breadboard wire** should bring up both types.

If you are still uncomfortable with breadboarding circuits, search the internet for additional tutorials and perhaps even simple projects you can use for practice. When you are ready, return to Chapter 3 and start building the electronics for your robot.

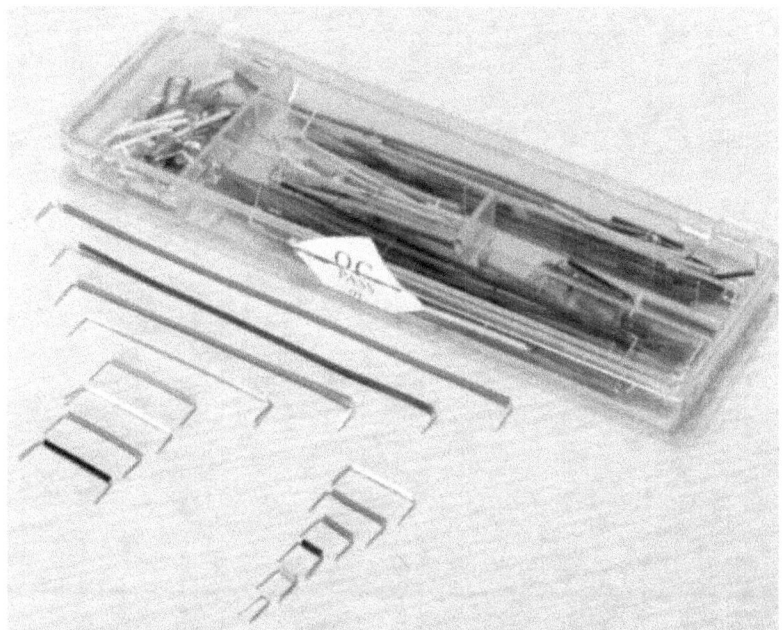

Figure A.5: Wire kits make it easy to breadboard circuits.

Figure A.6: Special jumpers can also be used on breadboards.

www.ingramcontent.com/pod-product-compliance
Lightning Source LLC
Chambersburg PA
CBHW082353220526
45470CB00008B/2725